Gordon J. Hartley
March 1979.

41071

AF681209

CAMBRIDGE CLASSICAL TEXTS AND COMMENTARIES

EDITORS

C. O. BRINK D. W. LUCAS F. H. SANDBACH

4

CICERO'S LETTERS TO ATTICUS

VOLUME II

CICERO'S LETTERS TO ATTICUS

EDITED BY

D. R. SHACKLETON BAILEY

VOLUME II

58–54 B.C.

46–93 (BOOKS III AND IV)

CAMBRIDGE

AT THE UNIVERSITY PRESS

1965

PUBLISHED BY
THE SYNDICS OF THE CAMBRIDGE UNIVERSITY PRESS

Bentley House, 200 Euston Road, London, N.W. 1
American Branch: 32 East 57th Street, New York, N.Y. 10022
West African Office: P.O. Box 33, Ibadan, Nigeria

©
CAMBRIDGE UNIVERSITY PRESS
1965

Printed in Great Britain at the University Printing House, Cambridge
(Brooke Crutchley, University Printer)

LIBRARY OF CONGRESS CATALOGUE
CARD NUMBER: 65-18929

CONTENTS

CONTENTS

ABBREVIATIONS

The following may be noted:

Broughton = T. R. S. Broughton, *The Magistrates of the Roman Republic* (New York, 1951–60). References unless otherwise stated are to Vol. II.

Drumann–Groebe = K. W. Drumann and P. Groebe, *Geschichte Roms.* 2nd edition (Leipzig, 1899–1929).

K.–S. = R. Kühner and C. Stegmann, *Ausführliche Grammatik der lateinischen Sprache*: Satzlehre. 3rd edition (Leverkusen, 1955).

Lehmann, *de Cic. ad Att.* = C. A. Lehmann, *De Ciceronis ad Atticum epistulis recensendis et emendandis* (Berlin, 1892).

Lehmann, *Quaest.* = C. A. Lehmann, *Quaestiones Tullianae* (Prague and Leipzig, 1886).

L.–S.–J. = Liddell–Scott–Jones, Greek–English Lexicon, 9th edition.

Mommsen, *St.* = Th. Mommsen, *Römisches Staatsrecht*, 3rd edition (Leipzig, 1887–8).

Mommsen, *Str.* = Th. Mommsen, *Römisches Strafrecht* (Leipzig, 1899).

Otto, *Sprichwörter* = A. Otto, *Die Sprichwörter und sprichwörtlichen Redensarten der Römer* (Leipzig, 1890).

Propertiana = D. R. Shackleton Bailey, *Propertiana* (Cambridge, 1956).

RE. = Pauly–Wissowa, *Realencyklopädie.*

Sjögren, *Comm. Tull.* = H. Sjögren, *Commentationes Tullianae* (Uppsala, 1910).

Thes. = *Thesaurus Linguae Latinae.*

ABBREVIATIONS

Towards a Text = D. R. Shackleton Bailey, *Towards a Text* of Cicero, *ad Atticum* (Cambridge, 1960).

T.–P. = R. Y. Tyrell and L. C. Purser, *The Correspondence of Cicero*, third and second editions (Dublin, 1904–33).

The Letters to Quintus and to M. Brutus are cited from W. S. Watt's Oxford Text, Asconius by line and page from that of A. C. Clark.

TEXT AND TRANSLATION

46–93 (BOOKS III–IV)

58–54 B.C.

INDEX SIGLORVM

E = Ambrosianus E 14 inf. (saec. XIV).
G = Parisinus 'Nouv. Fonds' 16248 (saec. XIV–XV).
H = Landianus 8 (saec. XIV–XV).
N = Laurentianus (ex Conv. Suppr.) 49 (saec. XIV–XV).
V = Palatinus Lat. 1510 (saec. XV).
O = Taurinensis Lat. 495 (saec. XV).
R = Parisinus Lat. 8538 (anno 1419 scriptus).
P = Parisinus Lat. 8536 (saec. XV).
Ant. = Antonianus a Malaespina citatus.
F = Faërni codex a Malaespina vel Vrsino citatus.
codd. Mal. = Malaespinae codices ab ipso generatim citati.
Σ = consensus codicum EGNVR vel omnium vel eorum qui quoque loco praesto sunt.
M = Mediceus 49. 18 (anno 1393 scriptus).
b = Berolinensis (ex bibl. Hamiltoniana) 168 (saec. XV).
d = Laurentianus (ex bibl. aedilium) 217 (saec. XV).
m = Berolinensis (ex bibl. Hamiltoniana) 166 (anno 1408 scriptus).
s = Vrbinas 322 (saec. XV).
δ = consensus codicum *bdms*.
Δ = consensus codicis M cum codicibus *bdms*, aut omnibus aut tribus.
Ω = consensus codicum Σ et M = archetypum omnium quos supra nominavi codicum.
C = lectiones margini editionis Cratandrinae (an. 1528) adscriptae.
c = lectiones in textu eiusdem editionis primum prolatae.
Z = Tornesianus (de Z^l, $Z^{(l)}$, Z^b, $Z^{(b)}$, Z^β, $Z^{(\beta)}$, Z^t, $Z^{(t)}$ vide Vol. I, p. 93).
λ = lectiones in margine alterius editionis Lambinianae (an. 1572–3) veteri codici (v.c.) adtributae.
Lamb. (*marg.*) = lectiones ibidem quibus littera 'L.' praefixa est.
ς = lectiones ex codicibus deterioribus, ut videntur, hic illic citatae vel in editionibus ante Cratandrinam primum inventae.
M^2, O^2, etc. = codicum M, O, etc. secundae manus.

46 (III. 1)

Scr. in exsilium proficiscens c. xi Kal. Apr., ut vid., an. 58

CICERO ATTICO SAL.

Cum antea maxime nostra interesse arbitrabar te esse nobiscum, tum vero, ut legi rogationem, intellexi ad iter id quod constitui nihil mihi optatius cadere posse quam ut tu me quam primum consequare, ut, cum ex Italia profecti essemus, sive per Epirum iter esset faciendum tuo tuorumque praesidio uteremur, sive aliud quid agendum esset certum consilium de tua sententia capere possemus. quam ob rem te oro des operam ut me statim consequare; quod eo facilius potes, quoniam de provincia Macedonia perlata lex est. pluribus verbis tecum agerem nisi pro me apud te res ipsa loqueretur.

47 (III. 3)

Scr. in itinere c. ix Kal. Apr., ut vid., an. 58

CICERO ATTICO SAL.

Vtinam illum diem videam cum tibi agam gratias quod me vivere coëgisti! adhuc quidem valde me paenitet. sed te oro ut ad me Vibonem statim venias, quo ego multis de causis converti iter meum. sed eo si veneris, de toto itinere ac fuga mea consilium capere potero. si id non feceris, mirabor; sed confido te esse facturum.

Ep. 46] 8 quod eo *c*: *om.* Ω.
Ep. 47] 2 quidem *V*: eq- *GNORΔ*.
3 de *om. Mm.*
6 cura ut cito (cito ut *R*) advoles *add. in fine VR.*

46 (III. 1)

Place unknown, c. 22 March (?) 58

CICERO TO ATTICUS

Even before I read the bill I felt it was very important to me to have you with me, and now that I have done so I see that nothing could be more desirable with a view to the journey on which I have decided than that you should overtake me as soon as possible, so that when I leave Italy I may have the protection of you and your people if I am to travel through Epirus, or if I am to take some different course I may have your advice in fixing upon a plan. So I beg you to try to overtake me directly. You can do so more easily now that the law about the province of Macedonia has gone through. I should urge you at greater length, but you have the facts and they will speak for me.

47 (III. 3)

En route, c. 24 (?) March 58

CICERO TO ATTICUS

I hope I may see the day when I shall thank you for making me go on living. So far I am heartily sorry you did. But I beg you to come to me as soon as possible at Vibo, where I am going. I have changed direction for many reasons. But if you come there, I shall be able to make a plan for my whole journey and exile. If you do not do that, I shall be surprised; but I am confident you will.

48 (III. 2)

Scr. Naribus Lucanis vi Kal. Apr. an. 58

CICERO ATTICO SAL.

Itineris nostri causa fuit quod non habebam locum ubi pro meo iure diutius esse possem quam fundum Siccae, praesertim nondum rogatione correcta; et simul intellegebam ex eo loco, si te haberem, posse me Brundisium referre, sine te autem non esse nobis illas partes tenendas propter Autronium. nunc, ut ad te antea scripsi, si ad nos veneris, consilium totius rei capiemus. iter esse molestum scio, sed tota calamitas omnis molestias habet. plura scribere non possum; ita sum animo perculso et abiecto. cura ut valeas. data VI Kal. Apr. Narib. Luc.

49 (III. 4)

Scr. Vibone, ut vid., iii Non. Apr. an. 58

CICERO ATTICO SAL.

Miseriae nostrae potius velim quam inconstantiae tribuas quod a Vibone quo te arcessebamus subito discessimus. adlata est enim nobis rogatio de pernicie mea, in qua quod correctum esse audieramus erat eius modi ut mihi ultra quadringenta milia liceret esse, illuc pervenire non liceret. statim iter Brundisium versus contuli ante diem rogationis, ne et Sicca, apud quem eram, periret et quod Melitae esse non licebat. nunc tu propera ut nos consequare, si modo recipiemur. adhuc invitamur benigne, sed quod superest timemus. me, mi Pomponi,

---

Ep. 48] 2 fundum *C*: in f- Ωλ.
3–4 si te haberem *C*: se iter habere (si...haberem *NO*) Σ: si recte habere *m* (*de M non liquet*).
5 Autronium *vulg.*: antro- *GNMdm*: anto- *HVRbs*.
9 Kal. *scripsi*: Id. Ω. Narib. Luc. *Klotz*: naris luc. Ω*Z*.
Ep. 49] 3 correctum *C*: confe- Ω.

48 (III. 2)

Nares Lucanae, 27 March 58

CICERO TO ATTICUS

The reason for my route was that I have no place where I can feel free to stay longer than at Sicca's farm, especially as the bill has not yet been amended. Also I have in mind that if I have you with me I can get back from there to Brundisium, while without you I had best not make for those parts on account of Autronius. Now, as I wrote to you previously, if you join me we shall make a comprehensive plan. I know the journey is troublesome, but the whole disaster is full of all manner of troubles. I cannot write any more, I am too stricken and dejected. Dispatched 27 March, Nares Lucanae.

49 (III. 4)

Vibo(?), 3 April 58

CICERO TO ATTICUS

I hope you will put it down to the misery I am in rather than to fickleness of purpose that I have suddenly left Vibo after I have been asking you to come there. The bill for my destruction has come into my hands, and the correction about which I had heard was to the effect that I am allowed to live at a distance of not less than 400 miles, but am not allowed to get there. I at once changed my course for Brundisium before the bill should become law for fear that my host Sicca might be ruined too and because I cannot stay in Malta. Do make haste to catch me up, that is if I can get anyone to take me in. So far people invite me kindly enough, but I am afraid of what is to

---

4 quadringenta *GOPΔ*: quadraginta *HVR*: XL *N*.
5 illuc *HVP*: illoc *GORMbd*: -o cum *ms*: illec *N*.

valde paenitet vivere, qua in re apud me tu plurimum valuisti. sed haec coram. fac modo ut venias.

50 (III. 5)

Scr. Thuriis viii Id. Apr. an. 58

CICERO ATTICO SAL.

Terentia tibi et saepe et maximas agit gratias. id est mihi gratissimum. ego vivo miserrimus et maximo dolore conficior. ad te quid scribam nescio. si enim es Romae, iam me adsequi non potes; sin es in via, cum eris me adsecutus, coram agemus quae erunt agenda. tantum te oro ut, quoniam me ipsum semper amasti, ut eodem amore sis; ego enim idem sum. inimici mei mea mihi, non me ipsum ademerunt. cura ut valeas. data VIII Id. Apr. Thuri.

51 (III. 6)

Scr. in Tarentino xiv Kal. Mai. an. 58

CICERO ATTICO SAL.

Non fuerat mihi dubium quin te Tarenti aut Brundisi visurus essem idque ad multa pertinuit, in eis et ut in Epiro consisteremus et de reliquis rebus tuo consilio uteremur. quoniam id non contigit, erit hoc quoque in magno numero nostrorum malorum. nobis iter est in Asiam, maxime Cyzicum. tibi meos commendo. me vix misereque sustento. data XIIII Kal. Mai. de Tarentino.

Ep. 50] 6 ut eodem *GNΔλ*: eodem *VR*.
Ep. 51] 2 in eis *NOR*: meis *HΔ*: me his *G*.
5–6 meos tibi Δ.
6 XIIII $\Omega Z^{(b)}$: VIIII *P*: XIII *c*.
7 de Tarentino *etiam Cλ*.

follow. For my part, my dear Pomponius, I am heartily sorry to be alive. In that decision you weighed with me most. But of this when we meet. Only do come.

50 (III. 5)

Thurii, 6 April 58

CICERO TO ATTICUS

Terentia tells me continually how very thankful she is to you. For that I am most grateful. My life is a misery and I am overwhelmed by profound unhappiness. What to write to you I don't know, for if you are in Rome you cannot catch up with me now, whereas if you are on your way we shall discuss what needs discussing together when you join me. All I beg of you is, since you have always loved me for my own sake, not to change. I am the same man. My enemies have robbed me of what I have, but not of what I am. Take care of your health. Dispatched 6 April, Thurii.

51 (III. 6)

Near Tarentum, 17 April 58

CICERO TO ATTICUS

I had counted on seeing you at Tarentum or Brundisium, and it was from many points of view desirable that I should, for example so that I could stop in Epirus and to have your advice about all other points. Since it has not happened I must include this in the multitude of my misfortunes. My way lies to Asia, Cyzicus for preference. I commend my family to your care. I keep going with difficulty and wretchedness. Dispatched 17 April, near Tarentum.

52 (III. 7)

Scr. Brundisi prid. Kal. Mai. an. 58 (§ 3)

CICERO ATTICO SAL.

1 Brundisium veni a. d. XIIII Kal. Mai. eo die pueri tui mihi a te litteras reddiderunt, et alii pueri post diem tertium eius diei alias litteras attulerunt. quod me rogas et hortaris ut apud te in Epiro sim, voluntas tua mihi valde grata est et minime nova. ⟨es⟩set consilium mihi quidem optatum si liceret ibi omne tempus consumere; odi enim celebritatem, fugio homines, lucem aspicere vix possum, esset mihi ista solitudo, praesertim tam familiari in loco, non amara. sed itineris causa ut deverterer, primum est devium, deinde ab Autronio et ceteris quadridui, deinde sine te. nam castellum munitum habitanti mihi prodesset, transeunti non est necessarium. quod si auderem, Athenas peterem. sane ita cadebat ut vellem. nunc et nostri hostes ibi sunt et te non habemus et veremur ne interpretentur illud quoque oppidum ab Italia non satis abesse, nec scribis quam ad diem te exspectemus.

2 Quod me ad vitam vocas, unum efficis ut a me manus abstineam, alterum non potes ut me non nostri consili vitaeque paeniteat. quid enim est quod me retineat, praesertim si spes ea non est quae nos proficiscentis prosequebatur? non faciam ut enumerem miserias omnis in quas incidi per summam iniuriam et scelus non tam inimicorum meorum quam invidorum, ne et meum maerorem exagitem et te in eundem luctum vocem; hoc adfirmo, neminem umquam tanta calamitate esse adfectum, nemini mortem magis optandum

---

Ep. 52] 1, 1 XIIII $\Sigma Z^{(b)}\lambda$: *om. Mm.* et eo *Mm.*
2 diei Ω: die *C.*
5 esset *Bentivolius*: sed Ω.
8 sed *V*: si *EN*: sit *GRΔ*: sic *m.*

52 (III. 7)

Brundisium, 29 April 58

CICERO TO ATTICUS

1 I arrived at Brundisium on 17 April. Your boys gave me a letter from you that day, and others brought me another letter two days later. I am deeply touched and not at all surprised by the kindness of your pressing invitation to stay at your place in Epirus. The idea would be much to my mind if I could spend all the time there—I hate crowds and shun my fellow creatures, I can hardly bear the light of day. The solitude, especially in such a friendly place, would be no hardship to me. But as a stopping-place *en route*, to begin with it's off my road, and then it's only four days' journey from Autronius and the others, and then you would not be there. A fortified place would have its advantages if I were going to live there, but if I am just passing through it isn't necessary. If I dared, I should make for Athens. Certainly as things were falling out I should have wanted to do so. But as it is, I have enemies there, I don't have you, and I am afraid they may make out that even Athens is not far enough away from Italy; and you don't say when I am to expect you.

2 Your exhortation to me to live is only partially effective. You keep me from laying violent hands upon myself, but you cannot keep me from regretting my decision and the fact that I am alive. What is there to hold me, especially if the hope which followed me when I left Rome is no more? I shall not proceed to make a catalogue of all the tribulations on which I have fallen by the signal injury and villainy not so much of those who hated me as of those who were jealous of me. That would only be to bring my grief to the surface and ask you to share my mourning. This I do say flatly, that no one ever suffered so crushing a blow or had greater cause to pray for

fuisse. cuius oppetendae tempus honestissimum praetermissum est; reliqua tempora sunt non iam ad medicinam sed ad finem doloris.

3 De re publica video te colligere omnia quae putes aliquam spem mihi posse adferre mutandarum rerum; quae quamquam exigua sunt, tamen, quoniam placet, exspectemus.

Tu nihilominus, si properaris, nos consequere. nam aut accedemus in Epirum aut tarde per Candaviam ibimus. dubitationem autem de Epiro non inconstantia nostra adferebat sed quod de fratre, ubi eum visuri essemus, nesciebamus; quem quidem ego nec ⟨quo⟩ modo visurus nec ut dimissurus sim scio. id est maximum et miserrimum mearum omnium miseriarum.

Ego et saepius ad te et plura scriberem nisi mihi dolor meus cum omnis partis mentis tum maxime huius generis facultatem ademisset. videre te cupio. cura ut valeas. d. prid. Kal. Mai. Brundisio proficiscens.

53 (III. 8)

Scr. Thessalonicae iv Kal. Iun. an. 58 (§ 4)

CICERO ATTICO SAL.

1 Scripseram ad te quas ob causas in Epirum non essemus profecti, quod et Achaia prope esset plena audacissimorum inimicorum et exitus difficilis haberet cum inde proficisceremur. accessit, cum Dyrrachi essemus, ut duo nuntii adferrentur, unus, classe fratrem Epheso Athenas, alter, pedibus per Macedoniam venire. itaque illi obviam misimus Athenas ut inde

---

2, 11 sunt non Σ*m*λ: non sunt Δ (*sed in M corr. manus prima, ut vid.*) Z^b (*nimirum aliud agens Bosius vulgatum sequebatur*). iam *Ant.* Z^b: tam Ωλ. sed *GNVR*ΔZ^bλ: quam *Es*.

3, 2 quamquam *Hms*: quam Ω (*om. R*).

5 accedemus *Es*: -damus Ω.

death. I might have met it with honour, but the moment was let pass. From the time that remains I do not look any longer for a remedy but only for an end to my misery.

3 I see you are collecting every item in the political news which you think could afford me some hope of a change. They don't amount to much, but, since you wish, let us wait and see.

If you make haste you will catch me up even now, as I shall either go to Epirus or travel slowly through Candavia. My hesitation about Epirus is not due to fickleness of purpose but to uncertainty about my brother, i.e. where I am to meet him—though how I am to meet him or to say good-bye to him I know not. That is the greatest and saddest of all my sorrows.

I should write to you more often and more fully, if it were not that my distress has robbed me of all my mental powers and more particularly of this sort of faculty. I long to see you. Take care of your health. Dispatched 29 April, on departure from Brundisium.

53 (III. 8)

Thessalonica, 29 May 58

CICERO TO ATTICUS

1 I have already written to you the reasons why I did not go to Epirus, namely the proximity of Achaea which is full of enemies who would stop at nothing and the difficulty of getting out of the country when I came to leave it. Furthermore, while I was at Dyrrachium two messages arrived, one to the effect that my brother was travelling by sea from Ephesus to Athens, the other that he was proceeding by land through Macedonia. I therefore sent a messenger to meet him at

---

8 quomodo ς: modo Ω. ut *Bosius*: vi *N*: cui *E*: vidi *G*: ubi *V*Δ: *nugas habent RP*.

9 sim *ms*: sum Ω. 13 d. *scripsi*: data Ω.

14 Brundisio *Klotz*: -sii Ω.

Thessalonicam veniret. ipsi processimus et Thessalonicam a. d. x Kal. Iun. venimus, neque de illius itinere quicquam certi habebamus nisi eum ab Epheso ante aliquanto profectum.

2 Nunc istic quid agatur magno opere timeo. quamquam tu altera epistula scribis Id. Mai. audiri fore ut acrius postularet⟨ur⟩, altera iam esse mitiora. sed haec est pridie data quam illa, quo conturbor magis. itaque cum meus me maeror cottidianus lacerat et conficit, tum vero haec addita cura vix mihi vitam reliquam facit.

Sed et navigatio perdifficilis fuit et ille incertus ubi ego essem fortasse alium cursum petivit. nam Phaëtho libertus eum non vidit; vento reiectus ab Ilio in Macedoniam Pellae mihi praesto fuit. reliqua quam mihi timenda sint video, nec quid scribam habeo et omnia timeo, nec tam miserum est quicquam quod non in nostram fortunam cadere videatur. equidem adhuc miser in maximis meis aerumnis et luctibus hoc metu adiecto maneo Thessalonicae suspensus nec audeo quicquam.

3 Nunc ad ea quae scripsisti. Tryphonem Caecilium non vidi. sermonem tuum et Pompei cognovi ex tuis litteris. motum in re publica non tantum ego impendere video quantum tu aut vides aut ad me consolandum adfers. Tigrane enim neglecto sublata sunt omnia. Varroni me iubes agere gratias. faciam; item Hypsaeo. quod suades ne longius discedamus dum acta mensis Mai ad nos perferantur, puto me ita esse facturum; sed ubi, nondum statui. atque ita perturbato sum animo de Quinto

---

Ep. 53] **2**, 1–2 tu altera M^{3}: tua a- *H* (-am) *bds*: tua littera (*vel* lirā) *GNORM*: tua *Vm*.

2 epistula *V*δ: -am *GNORM*. audiri Z^{b}λ: -re Ω. postularetur *Bosius*: -ret Ω: -rer *Ant.*

4 conturbor Σ*m*: -ber Δ.

8 Phaetho (*sed* pheton) libertus eum *Brunus* (M^{4}): phetolibeum *vel sim.* Ω.

9 Ilio *F*: illo Ω.

12 equidem *ERs*: q- *HNV*Δ: quid est *G*.

3, 3 impendere *om. Mm.*

8 perturbatus *EN*.

Athens, to tell him to go from there to Thessalonica. I myself went on and reached Thessalonica on 23 May, but I still have no definite information about his route except that he left Ephesus some time ago.

2 And now I am very anxious about what is happening in Rome. True, in one of your letters dated the Ides of May you write that according to report he will be vigorously prosecuted, whereas in another you say that the atmosphere has become milder. But that letter was dispatched the day before the other, which makes me doubly uneasy. So with this new anxiety on top of my own chronic grief that tears my heart and wears me down, I am pretty well at the end of my tether.

However, sailing was very difficult, and perhaps in uncertainty of my whereabouts he has taken a different direction. My freedman Phaetho did not see him. He was turned back to Macedonia from Ilium by a contrary wind and reported to me at Pella. Clearly I have every reason for apprehension, but I don't know what to write. I'm afraid of all manner of things, and nothing seems too bad to happen to us in our present plight. As for my unhappy self, plunged in woes and afflictions of my own and now with this added fear, I am staying for the present at Thessalonica, in suspense and not daring to make any move.

3 Now to answer the points in your letter. I have not seen Caecilius Trypho. I have taken note of your talk with Pompey as recounted in your letter. I perceive no sign of so great a movement in the political situation as you either see coming or profess to see in order to comfort me. If Tigranes has been passed over, it all goes by the board. You tell me to thank Varro; so I shall, and Hypsaeus likewise. I think I shall follow your advice not to go further away until the public transactions for May reach me. But where to go, I have not yet made up my mind. Indeed I am so worried about Quintus that I cannot

ut nihil queam statuere; sed tamen statim te faciam certiorem.

4 Ex epistularum mearum inconstantia puto te mentis meae motum videre, qui, etsi incredibili et singulari calamitate adflictus sum, tamen non tam est ex miseria quam ex culpae nostrae recordatione commotus. cuius enim scelere impulsi et proditi simus iam profecto vides; atque utinam ante vidisses neque totum animum tuum maerori mecum simul dedisses! qua re cum me adflictum et confectum luctu audies, existimato me stultitiae meae poenam ferre gravius quam eventi, quod ei crediderim quem esse nefarium non putarim. me et meorum malorum maeror et metus de fratre in scribendo impedit. tu ista omnia vide et guberna. Terentia tibi maximas gratias agit. litterarum exemplum quas ad Pompeium scripsi misi tibi. data IIII Kal. Iun. Thessalonicae.

54 (III. 9)

Scr. Thessalonicae Id. Iun. an. 58 (*§ 3*)

CICERO ATTICO SAL.

1 Quintus frater cum ex Asia discessisset ante Kal. Mai. et Athenas venisset Id. Mai., valde fuit ei properandum, ne quid absens acciperet calamitatis, si quis forte fuisset qui contentus nostris malis non esset. itaque eum malui properare Romam quam ad me venire; et simul (dicam enim quod verum est, ex quo magnitudinem mearum miseriarum perspicere possis) animum inducere non potui ut aut illum, amantissimum mei,

---

4, 1 constantia *Mm.*
3 est *om. ENV.*
5 simus *EGδ*: sumus *NVRM.* ante Σ: iam ante *HΔ.*
10 m(a)eror *Pbds*: memor *GNVRMm*: memor^{a} *H*: memoria *EO.*
12 tibi misi. vale *c.*
13 Iun(ias) Σ: Maias *Mm.*

make up my mind about anything. However, I shall let you know at once.

4 From the way my letters chop and change I expect you perceive the agitation of my mind, which is not so much caused by unhappiness, smitten though I am by a disaster beyond belief or example, as by the recollection of my own fault. Surely you see now whose villainy it was that urged me on and betrayed me. If only you had seen it sooner and not surrendered your whole soul to grief as I did myself! When therefore you hear of me as plunged in the lowest depths of misery, you are to suppose that the penalty of my folly weighs upon me more heavily than that of the event itself, in that I trusted a man whom I did not think to be a blackguard. Grief for my own troubles and fear on my brother's behalf alike make it hard for me to write. Please see to everything over there and direct. Terentia says she is most grateful to you. I am sending you a copy of a letter I have written to Pompey. Dispatched 29 May, Thessalonica.

54 (III. 9)

Thessalonica, 13 June 58

CICERO TO ATTICUS

1 My brother Quintus left Asia before the Kalends of May and reached Athens on the Ides. He had to make great haste for fear a blow might be struck against him in his absence—there might perhaps have been some for whom our present misfortunes are not enough. I therefore thought it better for him to hurry on to Rome rather than to come to me. At the same time (I must tell the truth, and from it you can gauge the extent of my wretchedness) I could not bring myself to meet him, devoted to me and soft-hearted as he is, in so sad a state,

Ep. 54] 1, 2 maiis *bds*: maias Ω.
6 mearum miseriarum *EGVR*: mis- me- *HNΔ*.

mollissimo animo, tanto in maerore adspicerem aut meas miserias luctu adflictus et perditam fortunam illi offerrem aut ab illo adspici paterer. atque etiam illud timebam, quod profecto accidisset, ne a me digredi non posset. versabatur mihi tempus illud ante oculos cum ille aut lictores dimitteret aut vi avelleretur ex complexu meo. huius acerbitatis eventum altera acerbitate non videndi fratris vitavi. in hunc me casum vos vivendi auctores impulistis. itaque mei peccati luo poenas.

2 quamquam me tuae litterae sustentant—e quibus quantum tu ipse speres facile perspicio; [et] quae quidem tamen aliquid habebant solaci ante quam eo venisti a Pompeio: 'nunc Hortensium adlice et eius modi viros.' obsecro, mi Pomponi, nondum perspicis quorum opera, quorum insidiis, quorum scelere perierimus? sed tecum haec coram agemus; tantum dico quod scire te puto: nos non inimici sed invidi perdiderunt. nunc si ita sunt quae speras, sustinebimus nos et spe qua iubes nitemur; sin, ut mihi videntur, ⟨in⟩firma sunt, quod optimo tempore facere non licuit minus idoneo fiet.

3 Terentia tibi saepe agit gratias. mihi etiam unum de malis in metu est, fratris miseri negotium; quod si sciam cuius modi sit, sciam quid agendum mihi sit. me etiam nunc istorum beneficiorum et litterarum exspectatio, ut tibi placet, Thessalonicae tenet. si quid erit novi adlatum, sciam de reliquo quid agendum sit. tu si, ut scribis, Kal. Iun. Roma profectus es, prope diem nos videbis. litteras quas ad Pompeium scripsi tibi misi. data Id. Iun. Thessalonicae.

1, 9 adflictus *Reid*: -tas Ω.
12 lictores *Gs*: lect- *vel sim. cett.*
13 eventum *ENORMm*: -tu *GVP*.
14 altera acerbitate *Em*: -ram -em *GNVR*Δ (*sed* aliam *bds*).

2, 2 quae ς: et q- Ω.
6 haec ΣZ^{l}: haec omnia Δ.
9 infirma ς: f- Ω.
10 tempore ς: genere (-ri *G*) Ω*C*λ.

nor yet, bowed in woe as I am, to thrust my misery and ruin in his way or allow him to see it. I was afraid too that he would not have the heart to leave me, nor would he, I am sure. I pictured the moment when he would either have to dismiss his lictors or be torn by force from my arms. This bitter consequence I have avoided by the bitterness of not seeing my brother. Such is the dilemma into which you and others who urged me to live on have pushed me. Well, I am paying for
2 my mistake. Not but what your letter gives me comfort, though I can easily see from it how much hope *you* have! But still it offered some consolation until you got to the point where, after mention of Pompey, you go on: 'And now draw in Hortensius and personages of that sort.' For mercy's sake, my dear fellow, don't you *yet* see whose treachery and villainy engineered my ruin? But I will take that up with you when we are together. I say only this, and I think you know it is so: it was not enemies but jealous friends who ruined me. As things are, if your hopes are well founded, I shall keep going and trust to the hope which you tell me to trust. If on the other hand they are unreliable, as they seem to me to be, then what I was not allowed to do at the right time shall be done at the wrong.

3 Terentia often says how grateful she is to you. For me too fear forms one among my troubles, I mean my unfortunate brother's affair. If I could know how that stands, I should know how to proceed. I am still kept at Thessalonica, as you advise, waiting for the good offices and letters you mention. When I get some news I shall be able to plan ahead. If, as you say, you left Rome on the Kalends of June, you will be seeing me soon. I am sending you a letter I have written to Pompey. Dispatched Ides of June, Thessalonica.

55 (III. 10)

Scr. Thessalonicae xiv Kal. Quint. an. 58 (§ 3)

CICERO ATTICO SAL.

1 Acta quae essent usque ad VIII Kal. Iun. cognovi ex tuis litteris; reliqua exspectabam, ut tibi placebat, Thessalonicae. quibus adlatis facilius statuere potuero ubi sim. nam si erit causa, si quid agetur, si spem videro, aut ibidem opperiar aut me ad te conferam; sin, ut tu scribis, ista evanuerint, aliquid aliud videbimus. omnino adhuc nihil mihi significatis nisi discordiam istorum; quae tamen inter eos de omnibus potius rebus est quam de me. itaque quid ea mihi prosit nescio. sed tamen quoad me vos sperare vultis, vobis obtemperabo.

2 Nam quod me tam saepe et tam vehementer obiurgas et animo infirmo esse dicis, quaeso, ecquod tantum malum est quod in mea calamitate non sit? ecquis umquam tam ex amplo statu, tam in bona causa, tantis facultatibus ingeni, consili, gratiae, tantis praesidiis bonorum omnium concidit? possum oblivisci qui fuerim? non sentire qui sim, quo caream honore, qua gloria, quibus liberis, quibus fortunis, quo fratre? quem ego, ut novum calamitatis genus attendas, cum pluris facerem quam me ipsum semperque fecissem, vitavi ne viderem, ne aut illius luctum squaloremque adspicerem aut ne me, quem ille florentissimum reliquerat, perditum illi adflictumque offerrem. mitto cetera intolerabilia; etenim fletu impedior. hic utrum tandem sum accusandus quod doleo an quod commisi ut haec aut non retinerem, quod facile fuisset nisi intra parietes meos de mea pernicie consilia inirentur, aut certe vivus non amitterem?

Ep. 55] 1, 5 aliquid aliud *V codd. Mal. Msmλ*: aliud aliq- *EGNR*.
2, 3 tam ex amplo ς: tam extemplo *NMm*: tam (*om. G*) extemplo a *EGVR*.
6 quid sim *G*.
10 ne me *GNVR*: me *EΔ*.
15 non *etiam* CZ^l.

55 (III. 10)

Thessalonica, 17 June 58

CICERO TO ATTICUS

1 Your letter has apprised me of public transactions up to 25 May. I await the rest, as you advised, at Thessalonica. When they arrive I shall be able to decide where to settle. If there's a movement, action, hope in view, I shall either stay where I am or move over to you. If on the other hand those prospects 'fade out', I shall look elsewhere. To be sure none of you tells me anything so far except the dissension among the powers that be, and as this is about everything under the sun rather than me I hardly see what good I get out of it. But still, as long as you all want me to hope, I shall comply.

2 When you take me to task, as you so often and so energetically do, and tell me I am lacking in fortitude, I should like to ask you whether there is any misfortune too great to have its place in this disaster of mine. Has any man ever fallen from so fine a position, with so good a cause, so strong in resources of talent, prudence, and influence, and in the support of all honest men? Can I forget what I was, or fail to feel what I am and what I have lost—rank, fame, children, fortune, brother? As to him, mark this novelty in affliction: loving him as I do and ever have done better than myself, I have avoided meeting him so as not to see him in the dress and grief of mourning nor yet to present myself—me, whom he left in the full tide of prosperity—before his eyes, a ruined and broken man. Of other intolerable tribulations I say nothing; indeed tears prevent me. In such circumstances am I really to be blamed for grieving, and not rather for my failure either to keep all these things, as I might easily have done if plans for my undoing had not been hatched under my own roof, or at least to lose them only with life?

3 Haec eo scripsi ut potius relevares me, quod facis, quam ut castigatione aut obiurgatione dignum putares, eoque ad te minus multa scribo quod et maerore impedior et quod exspectem istinc magis habeo quam quod ipse scribam. quae si erunt adlata, faciam te consili nostri certiorem. tu, ut adhuc fecisti, quam plurimis de rebus ad me velim scribas, ut prorsus ne quid ignorem. data XIIII Kal. Quint. Thessalonicae.

56 (III. 11)

Scr. Thessalonicae iv Kal. Quint. an. 58 (§ 2)

CICERO ATTICO SAL.

1 Me et tuae litterae et quidam boni nuntii, non optimis tamen auctoribus, et exspectatio vestrarum litterarum et quod tibi ita placuerat adhuc Thessalonicae tenebat. si accepero litteras quas exspecto, si spes erit ea quae rumoribus adferebatur, ad te me conferam; si non erit, faciam te certiorem quid egerim.

2 Tu me, ut facis, opera, consilio, gratia iuva. consolari iam desine, obiurgare vero noli; quod cum facis, ut ego tuum amorem et dolorem desidero! quem ita adfectum mea aerumna esse arbitror ut te ipsum consolari nemo possit. Quintum fratrem optimum humanissimumque sustenta. ad me obsecro te ut omnia certa perscribas. data IIII Kal. Quint.

57 (III. 12)

Scr. Thessalonicae xvi Kal. Sext. an. 58 (§ 3)

CICERO ATTICO SAL.

1 Tu quidem sedulo argumentaris quid sit sperandum et maxime per senatum, idemque caput rogationis proponi scribis qua re

---

Ep. 56] 1, 3 tenebat *ENVR*: -ant Gδ (*de M non liquet*).
Ep. 57] 1, 2 proponi *vulg.*: -is Ω.

3 I write this so that you may rather lighten my burden, as you do, than think me a proper subject for castigation or scolding; and I do not write more because I am prevented by grief on the one hand and on the other am in the position of waiting for news from Rome rather than having anything to communicate myself. When the news arrives I shall let you know my plans. Please write to me, as you have so far been doing, on as many points as you can, so that there may be *nothing* about which I am not informed. Dispatched 17 June, Thessalonica.

56 (III. 11)

Thessalonica, 27 June 58

CICERO TO ATTICUS

1 Your letter, certain favourable reports (not on the best of authority however), the expectation of letters from yourself and the others, and your earlier advice still keep me in Thessalonica. Once I receive the letters I am expecting, if the hope of which rumours reach me proves real, I shall move over to you; if not, I shall let you know what I do.

2 On your side, aid me, as you are doing, with effort, advice, and influence. Give up trying to comfort me, but don't scold. When you do this, how I miss your affection and sympathy! I believe you are affected by my misery to the point of being inconsolable yourself. Give a helping hand to Quintus, best and kindest of brothers. I beg you to send me full and reliable news on all matters. Dispatched 27 June.

57 (III. 12)

Thessalonica, 17 July 58

CICERO TO ATTICUS

1 You are at pains to argue what may be hoped for, particularly through the Senate, and yet you write that the clause in the bill

in senatu dici nihil liceat. itaque siletur. hic tu me accusas quod me adflictem, cum ita sim adflictus ut nemo umquam, quod tute intellegis. spem ostendis secundum comitia. quae ista est
2 eodem tribuno plebis et inimico consule designato? percussisti autem me etiam de oratione prolata. cui vulneri, ut scribis, medere, si quid potes. scripsi equidem olim iratus quod ille prior scripserat, sed ita compresseram ut numquam emanaturam putarem. quo modo exciderit nescio. sed quia numquam accidit ut cum eo verbo uno concertarem et quia scripta mihi videtur neglegentius quam ceterae puto ex se ⟨posse⟩ probari non esse meam. id, si putas me posse sanari, cures velim; sin plane perii, minus laboro.

3 Ego etiam nunc eodem in loco iaceo sine sermone ullo, sine cogitatione ulla. licet tibi, ut scribis, significarim ut ad me venires, id dono tamen et intellego te istic prodesse, hic ne verbo quidem levare me posse. non queo plura scribere nec est quod scribam; vestra magis exspecto. data XVI Kal. Sext. Thessalonicae.

58 (III. 14)

Scr. Thessalonicae xii Kal. Sext. an. 58 (§ *2*)

CICERO ATTICO SAL.

1 Ex tuis litteris plenus sum exspectatione de Pompeio, quidnam de nobis velit aut ostendat. comitia enim credo esse habita, quibus absolutis scribis illi placuisse agi de nobis. si tibi

---

2, 3 iratus *EVR*: ei ratus *GNMm*: ei ir- *Hbds*.
4–5 eman- put- *ENVR*: man- put- *Mm*: naturae imput- *G*.
5 exciderit *EV*: -int *NM*(?)*m*: -im *GR*.
7 ex se posse *scripsi*: ex se *EGV*: esse *ORMm*: posse *NP*.
3, 2 significarim *Victorius*: -rem Ω.
3 id dono tamen et *scripsi*, *similia iam Springer et Constans*: si donatam ut *NVRΔ*: si dona *E*: si donatam...posse *om*. *GH*.

which forbids any reference to the subject in the Senate is being posted up. So naturally silence reigns. And in these circumstances you take me to task for tormenting myself, when I am tried beyond any mortal that ever was, as you well know. You hold out hope after the elections. What does it amount to, with the same Tribune still in office and my enemy Consul-Desig-
2 nate? Then you have given me a blow about the circulation of that speech. Try, as you say, to patch up the damage if you can. I did write it long ago in a fit of annoyance with him because *he* had written against me, but I suppressed it and never expected it to leak out. How it did get out I don't know. However, as I have never exchanged a contentious word with him in my life and as it seems to me more carelessly written than my other compositions, I think it may be passed off on internal evidence as a forgery. Would you please see to that, if you think my case is curable? If I am past praying for, I don't so much care.

3 I am still stuck here, with no one to talk to and nothing to think about. I may have suggested to you, as you say, that you should join me; but I give that up, and realize that you are helping me where you are, whereas you could do nothing even verbally to lighten my load here. I can't write any more, nor have I anything to write about. I am rather waiting for news from your side. Dispatched 17 July, Thessalonica.

58 (III. 14)

Thessalonica, 21 July 58

CICERO TO ATTICUS

1 After your letter I am all expectation about Pompey, what he means by me or professes to mean. For I suppose that the elections are now over, at the conclusion of which you say he thought my case should be taken up. If you think me a fool to

stultus esse videor qui sperem, facio tuo iussu †et scio te meis epistulis† potius et meas spes solitum esse remorari. nunc velim mihi plane perscribas quid videas. scio nos nostris multis peccatis in hanc aerumnam incidisse. ea si qui casus aliqua ex parte correxerit, minus moleste feremus nos vixisse et adhuc vivere.

2 Ego propter viae celebritatem et cottidianam exspectationem rerum novarum non commovi me adhuc Thessalonica. sed iam extrudimur, non a Plancio (nam is quidem retinet) verum ab ipso loco minime apposito ad tolerandam in tanto luctu calamitatem. in Epirum ideo, ut scripseram, non veni quod subito mihi universi nuntii venerant et litterae qua re nihil esset necesse quam proxime Italiam esse. hinc, si aliquid a comitiis audierimus, nos in Asiam convertemus; neque adhuc stabat quo potissimum, sed scies. data XII Kal. Sext. Thessalonicae.

59 (III. 13)

Scr. Thessalonicae Non. Sext. an. 58 (§ 2)

CICERO ATTICO SAL.

1 Quod ad te scripseram me in Epiro futurum, postea quam extenuari spem nostram et evanescere vidi, mutavi consilium nec me Thessalonica commovi, ubi esse statueram quoad aliquid ad me de eo scriberes quod proximis litteris scripseras, fore uti secundum comitia aliquid de nobis in senatu ageretur; id tibi Pompeium dixisse. qua de re quoniam comitia habita sunt tuque nihil ad me scribis, proinde habebo ac si scripsisses

---

Ep. 58] 1, 4 etsi scio *F*.
5 meas spes *ms*: mea spes *Md*: mea spe Σ*b*. remorari *EH*: -re *GNVR* (remem-) Δ.
7 qui Σδ: quis *M*.
8 correxerit *NR*: -int *EV*Δ: correxit *G*.
2, 5 veni *HPZ*[b]λ: mei *R*: ivi *GN*: ii *EV*Δ.
6 subito *Z*[b]λ: -ti Ω. universi *etiam* λ.

hope, well, I do so at your bidding, and you (?) know that in your letters you have rather been wont to lead me on and keep my hopes alive (?). Now I should be glad if you would let me know the prospects fully and frankly, as you see them. I know that I have fallen into this unhappy plight by my own mistakes, which have been many. If chance in some measure puts them right I shall be less inclined to regret that I was ever born and that I am alive to-day.

2 I have not yet moved from Thessalonica because the route is so frequented and because I am waiting from day to day for news of a turn in the situation. But I am now being forced out, not by Plancius, who for his part would like me to stay on, but by the place itself—there could be no less suitable spot in which to bear calamity in such a state of grief as I am in. I have not gone to Epirus, as I wrote that I should, because all at once my letters and reports indicated with one accord that there was no occasion for me to be as near as possible to Italy. From here, once I hear something after the elections, I shall betake myself to Asia, where in particular I am not yet sure, but you shall know. Dispatched 21 July, Thessalonica.

59 (III. 13)

Thessalonica, 5 August 58

CICERO TO ATTICUS

1 Though I wrote to you that I should stay in Epirus, I changed my mind when I saw that my chances were thinning and fading away, and have not moved from Thessalonica. I had decided to stay there until I heard something from you about what you said in your last letter, namely that some move would be made about my case in the Senate after the elections, Pompey having told you so. Well, the elections are over and you don't write a word to me, so I shall regard that as equivalent to

nihil esse, meque temporis non longinqui spe ductum esse ⟨non⟩ moleste feram. quem autem motum te videre scripseras qui nobis utilis fore videretur, eum nuntiant qui veniunt nullum fore. in tribunis pl. designatis reliqua spes est. quam si exspectaro, non erit quod putes me causae meae, voluntati meorum defuisse.

2 Quod me saepe accusas cur hunc meum casum tam graviter feram, debes ignoscere, cum ita me adflictum videas ut neminem umquam nec videris nec audieris. nam quod scribis te audire me etiam mentis errore ex dolore adfici, mihi vero mens integra est. atque utinam tam in periculo fuisset, cum ego iis quibus meam salutem carissimam esse arbitrabar, inimicissimis crudelissimisque usus sum! qui ut me paulum inclinari timore viderunt, sic impulerunt ut omni suo scelere et perfidia abuterentur ad exitium meum.

Nunc quoniam iam est Cyzicum nobis eundum, quo rarius ad me litterae perferentur, hoc velim diligentius omnia quae putaris me scire opus esse perscribas. Quintum, fratrem meum, fac diligas. quem ego miser si incolumem relinquo, non me totum perisse arbitrabor. data Non. Sext.

60 (III. 15)

Scr. Thessalonicae xiv Kal. Sept. an. 58 (§8)

CICERO ATTICO SAL.

1 Accepi Id. Sext. quattuor epistulas a te missas: unam qua me obiurgas et rogas ut sim firmior; alteram qua Crassi libertum

---

Ep. 59] 1, 8 meque Ω: neque *P.* non *add. Otto.*
12 ac *post* meae *add.* ς (‘*veteres codices non habent copulationem* ac’ *marg. ed. alt. Lamb.*).
2, 7 paululum *G.*
9 exitium *GVs*: -tum *ERΔ*: exercitum *N.*
10 iam *EGV*: a me *R*: *om. NΔ.*
11 perferentur ς: -runtur Ω.

writing that there is nothing doing, and I shall not take it too hard that for no very long time I have been amused by hope. As for the change likely to turn to my advantage which you wrote that you saw coming, those who come from Rome say that nothing of the kind will happen. The remaining hope lies in the Tribunes-Designate. If I wait for that, you will have no reason to think that I have failed in what I owe to my cause and my friends' good will.

2 You repeatedly take me to task for bearing what has happened to me so hard, but you ought to forgive me when you see how I am tried beyond anything you ever witnessed or heard of. You say you hear that grief has actually disturbed the balance of my mind. No, my mind is sound enough. If it had only been equally so in the hour of danger, when I experienced the cruellest malice from those who I thought cared most about my welfare! The moment they saw me a little unnerved and inclined to waver, they pushed me on, employing all their wickedness and perfidy to bring about my downfall.

As matters stand I must go to Cyzicus, where letters will reach me less often. So I hope you will be all the more careful to send me full reports of everything you think I ought to know. Be good to my brother Quintus. If I leave him out of danger (ah me!), I shall not feel I am altogether lost. Dispatched Nones of August.

60 (III. 15)

Thessalonica, 17 August 58

CICERO TO ATTICUS

1 On the Ides of August I received four letters from you. In the first you scold me and ask me to show more fortitude. In the second you say that Crassus' freedman has told you how

Ep. 60] I, 2 et rogas *om.* Δ.

ais tibi de mea sollicitudine macieque narrasse; tertiam qua demonstras acta in senatu; quartam de eo quod a Varrone scribis tibi esse confirmatum de voluntate Pompei.

2 Ad primam tibi hoc scribo, me ita dolere ut non modo a mente non deserar sed id ipsum doleam, me tam firma mente ubi utar et quibuscum non habere. nam si tu me uno non sine maerore cares, quid me censes, qui et te et omnibus? et si tu incolumis me requiris, [et] quo modo a me ipsam incolumitatem desiderari putas? nolo commemorare quibus rebus sim spoliatus, non solum quia non ignoras sed etiam ne scindam ipse dolorem meum; hoc confirmo, neque tantis bonis esse privatum quemquam neque in tantas miserias incidisse. dies autem non modo non levat luctum hunc sed etiam auget. nam ceteri dolores mitigantur vetustate, hic non potest non et sensu praesentis miseriae et recordatione praeteritae vitae cottidie augeri. desidero enim non mea solum neque meos sed me ipsum. quid enim sum? sed non faciam ut aut tuum animum angam querelis aut meis vulneribus saepius manus adferam. nam quod purgas eos quos ego mihi scripsi invidisse et in eis Catonem, ego vero tantum illum puto ab isto scelere afuisse ut maxime doleam plus apud me simulationem aliorum quam istius fidem valuisse. ceteros quod purgas, debent mihi probati
3 esse tibi si sunt. sed haec sero agimus.

Crassi libertum nihil puto sincere locutum. in senatu rem probe scribis actam. sed quid Curio? an illam orationem non legit? quae unde sit prolata nescio. sed Axius eiusdem diei

2, 5 quo *P*: et quo Ωλ.
10 non modo levat *c*.
16 ego mihi *ERMm*: mihi *GNV*. scripsi *Em*: scribis *GNVRM(?)*.
18 aliorum Σ*C*: malo- Δ.
19 quod $Z^l\lambda$: quos Ω. probati *EGNVPZ*: purgati *HRMm*.
3, 4 axius *NC*: anx- *EGVR*: aux- *M(?)m*.

anxious he found me and how I have lost weight. In the third you give an account of transactions in the Senate. In the fourth you write that Varro has given you confirmation of Pompey's good will.

2 In reply to the first, I have this to say. My distress has not affected my mind; on the contrary one of its elements is that, my faculties being as sound as they are, I have neither matter nor company in which to use them. If the loss of my single self causes you some pain, what do you think *my* pain must be in losing you and all besides? And if you, who still enjoy your status in the community, feel my absence, how do you think *I* feel the loss of that very status? I don't propose to rehearse all the good things of which I have been stripped, for you know them already and I don't want to tear my own sore. But I do assert that no man has ever lost so much or fallen into such a pit of misery. Time, far from relieving this heartache, actually increases it. Other hurts grow less acute as they grow older, this cannot but increase from day to day from the sense of present misery and the recollection of the life that is past. I mourn the loss not only of the things and persons that were mine, but of my very self. What am I now? But I must refrain from torturing you with complaints or putting my hands too often to my wounds. As for your exculpation of those who I said were jealous of me, among whom you include Cato, I am so far from thinking him guilty of the villainy in question that it is one of my principal regrets that the pretences of others counted more with me than his good faith. As for the
3 rest whom you exculpate, I must approve them if you do. But it's late in the day for us to talk of this.

I imagine that Crassus' freedman spoke only in malice. You say that the debate in the Senate went well. But what of Curio? Or has he not read that speech? How it became public property I can't think. But Axius, in giving me an account of

scribens ad me acta non ita laudat Curionem. at potest ille aliquid praetermittere; tu, nisi quod erat, profecto non scripsisti. Varronis sermo facit exspectationem Caesaris. atque utinam ipse Varro incumbat in causam! quod profecto cum sua sponte tum te instante faciet.

4 Ego, si me aliquando vestri et patriae compotem fortuna fecerit, certe efficiam ut maxime laetere unus ex omnibus amicis meaque officia et studia quae parum antea luxerunt (fatendum est enim) sic exsequar ut me aeque tibi ac fratri et liberis nostris restitutum putes. si quid in te peccavi, ac potius quoniam peccavi, ignosce; in me enim ipsum peccavi vehementius. neque haec eo scribo quo te non meo casu maximo dolore esse adfectum sciam; sed profecto ⟨si⟩ quantum me amas et amasti tantum amare deberes ac debuisses, numquam esses passus me, quo tu abundabas, egere consilio, nec esses passus mihi persuaderi utile nobis esse legem de collegiis perferri. sed tu tantum lacrimas praebuisti dolori meo, quod erat amoris, tamquam ipse ego; quod meritis meis perfectum potuit, ut dies ac noctes quid mihi faciendum esset cogitares, id abs te meo, non tuo scelere praetermissum est. quod si non modo tu sed quisquam fuisset qui me Pompei minus liberali responso perterritum a turpissimo consilio revocaret, quod unus tu facere maxime potuisti, ⟨aut honeste occubuissemus⟩ aut victores hodie viveremus. hic mihi ignosces. me enim ipsum multo magis accuso, deinde te quasi me alterum, et simul meae culpae socium quae⟨ro⟩. ac si restituor, etiam minus videbimur

3, 5 at ς: ac Ω.
9 faciet Gδ: -iat *ENORM*: facit *V*.
4, 2 l(a)etere *V*δ: lat- Ω (*praet. V*).
8 si *Pm*: *om.* Ω.
11 perferri *GVRC*: prof- *ENMm*.
13 tamquam...praesto esses (7, 3) *om. GH*.
14 cogitares *P*: -em *NV*δ: -e *RM*(?). id ς: si Ω.
18 aut honeste occubuissemus *Watt* (aut occ- hon- *Corradus*): *om.* ΩZ^l.
21 quaero. ac ς: qu(a)e ac *VR*Δ (ac *etiam* λ): quare *P*: ac *N*.

the same day's proceedings, doesn't praise Curio so highly. However, he may be leaving something out, whereas you of course will not have written anything but what was fact. What Varro says gives some hope of Caesar. If only Varro himself would put his shoulder to the wheel! No doubt he will, under pressure from you as well as of his own volition.

4 If fortune ever gives me back my friends and my country, I promise to see that none of them has so much cause to rejoice as yourself. I shall be so assiduous in friendly offices and tokens of good will, which in the past I must confess have not been sufficiently conspicuous, that you will feel that I am restored to *you* no less than to my brother and our children. If I have treated you badly in any way, or rather *since* I have done so, forgive me. I have treated myself worse. If I write in this strain, it is not that I am unaware how deeply distressed you are by my misfortune. But surely, if now and in the past I had really deserved the affection you felt and feel for me, you would never have let me go short of that good advice of which you have so ample a store, never have let me be persuaded that the passage of the law about the clubs was to my advantage. But you had only tears for my distress, the tribute of affection, just as I had myself. Desert on my part might have gained me something beyond, your daily and nightly meditation as to what it was best for me to do; that, through my delinquency not yours, you did not furnish. If you, or anyone else for that matter, at the time when I was thrown into consternation by Pompey's ungenerous response had held me back from a most discreditable resolution, as you were uniquely in a position to do, I should either have met an honourable death or be living triumphant to-day. You must forgive me here. I am reproaching myself far more than you, and if I do reproach you it is as my *alter ego*; also I am looking for someone to share the blame. If I am restored, my fault will come to seem less grave,

deliquisse abs teque certe, quoniam nullo nostro, tuo ipsius beneficio diligemur.

5 Quod te cum Culleone scribis de privilegio locutum, est aliquid, sed multo est melius abrogari. si enim nemo impediet, quid est firmius? sin erit qui ferri non sinat, idem senatus consulto intercedet. nec quicquam aliud opus est [quam] abrogari; nam prior lex nos nihil laedebat. quam si, ut est promulgata, laudare voluissemus aut, ut erat neglegenda, neglegere, nocere omnino nobis non potuisset. hic mihi primum meum consilium defuit, sed etiam obfuit. caeci, caeci inquam, fuimus in vestitu mutando, in populo rogando, quod, nisi nominatim mecum agi coeptum esset, fieri perniciosum fuit. sed pergo praeterita, verum tamen ob hanc causam, ut, si quid agetur, legem illam, in qua popularia multa sunt, ne tangatis.

6 Verum est stultum me praecipere quid agatis aut quo modo. utinam modo agatur aliquid! in quo ipso multa occultant tuae litterae, credo ne vehementius desperatione perturber. quid enim vides agi posse aut quo modo? per senatumne? at tute scripsisti ad me quoddam caput legis Clodium in curiae poste fixisse, ne referri neve dici liceret. quo modo igitur Domitius se dixit relaturum? quo modo autem iis quos tu scribis et de re dicentibus et ut referretur postulantibus Clodius tacuit? ac si per populum, poteritne nisi de omnium tribunorum pl. sententia? quid de bonis? quid de domo? poteritne restitui? aut si non poterit, egomet quo modo potero? haec nisi vides expediri, quam in spem me vocas? sin autem spei nihil est,

---

5, 3 quid CZ^bλ: sic Ω. firmius Δ: -mus Σ*d*. idem *Hervagius*: eidem Ω.
4 quam *del. Tunstall.*
7–8 meum...obfuit *etiam* λ.
11 agetur *Lambinus*: ageretur Ω.

6, 2 in quo ipso Z^b: in quo ipsa Σ: quo ipsa Δ.
4 at tute *bds*: astute *VR*: aut t- *N*: ast tute *M* (*sed* ast *in ras.*) *m*.
7 iis quos *s*: i(i)s (his, id) quo(d) Ω. 8 ac *etiam* λ.

and you at any rate will care for me, for the sake of what you have done for me since I have done nothing for you.

5 You say you have talked to Culleo about its being a law *ad hominem*. There is something in that, but it would be far better to get it repealed. That is the safest way if nobody obstructs. On the other hand, if anyone is going to block a law, he will equally veto a senatorial decree. And there is no need for anything else to be repealed, since the first law did not touch me. If I had chosen to welcome it when it was first promulgated or to ignore it as it deserved, it could have done me no harm whatsoever. That was the first point where my judgement let me down, or rather did me down. We were blind, yes, blind, to put on mourning and appeal to the people—a fatal blunder where no proceedings had been taken against me by name. But this is crying over spilt milk. I do it however for a reason, to dissuade you, if any move is to be made, from touching that law, which has many popular features.

6 But it is fatuous for me to lay down what you people are to do or how you are to do it. If only *something* is done! In that very regard your letters leave many things obscure, I suppose because you are afraid of despair throwing me quite off my balance. For what in your view can be done and how? Through the Senate? But you yourself told me that Clodius posted in the doorway of the Senate House a clause in his law banning 'any motion or mention'. How then could Domitius say that he would make a motion? And how was it that Clodius sat mum while the persons you mention spoke on the subject and demanded that a motion be put to the House? And if it is to go through the Assembly, will that be possible without the unanimous approval of the Tribunes? What about my property, and my house? Can it be restored, and, if not, how can I be? Unless you see solutions to these problems, what sort of a hope are you asking me to entertain? If on the other

quae est mihi vita? itaque exspecto Thessalonicae acta Kal. Sext., ex quibus statuam in tuosne agros confugiam, ut neque videam homines quos nolim et te, ut scribis, videam et propius sim si quid agatur, id quod intellexi cum tibi tum Quinto fratri placere, an abeam Cyzicum.

7 Nunc, Pomponi, quoniam nihil impertisti tuae prudentiae ad salutem meam, quod aut in me ipso satis esse consili decreras aut te nihil plus mihi debere quam ut praesto esses, quoniamque ego proditus, inductus, coniectus in fraudem omnia mea praesidia neglexi, totam Italiam [in me] erectam ad me defendendum destitui et reliqui, me, meos tradidi inimicis inspectante et tacente te, qui, si non plus ingenio valebas quam ego, certe timebas minus: si potes, erige adflictos et in eo nos iuva; sin omnia sunt obstructa, id ipsum fac ut sciamus et nos aliquando aut obiurgare aut communiter consolari desine. ego si tuam fidem accusarem, non me potissimum tuis tectis crederem; meam amentiam accuso, quod me a te tantum amari quantum ego vellem putavi. quod si fuisset, fidem eandem, curam maiorem adhibuisses, me certe ad exitium praecipitantem retinuisses, istos labores, quos nunc in naufragiis nostris suscipis, non subisses.

8 Qua re fac ut omnia ad me perspecta et explorata perscribas meque, ut facis, velis esse aliquem, quoniam qui fui et qui esse potui, iam esse non possum, et ut his litteris non te sed me ipsum a me esse accusatum putes. si qui erunt quibus putes opus esse meo nomine litteras dari, velim conscribas curesque dandas. data XIIII Kal. Sept.

6, 16 id quod *Klotz*: idque Ω.
7, 2 decreras *etiam* λ.
5 in me *secl. Ernesti.*
6 meos *ENOR*: meis *Mm*: meosque *GV*: meos me λ.
12 me *c*: *om.* Ω.
8, 6 XIII *HNc.*

hand there *is* no hope, what has life to offer me? So I am waiting at Thessalonica for the proceedings of the Kalends of August, in the light of which I shall decide whether to take refuge on your land, where I shall not see anyone I don't want and where I shall see you, so you say, and be closer if anything is done, as I understand you and my brother Quintus both think advisable, or go away to Cyzicus.

7 Now, Pomponius, I appeal to you. You did not give me of your worldly wisdom to save me from ruin, considering perhaps that I was competent to judge for myself or else that you owed me nothing more than your presence. Betrayed, decoyed, trapped as I was, I neglected all my defences, deserted and abandoned the Italy which stood as one man ready to protect me, and surrendered myself and my family to my enemies, while you, my superior in *sang-froid* if not in mother-wit, looked on and said nothing. Now, if it is in your power, raise me from the dust and help me in that. But if all paths are blocked, then tell me so, and don't let us have any more either of scolding or of consolatory commonplace. If I were saying anything against your good faith I should not choose your house for a refuge. It is my own folly that I reproach, in imagining that you felt as warmly towards me as I could have wished. Had that been so you would have brought not better faith indeed but greater solicitude to bear, you would certainly have held me back as I rushed upon destruction, and would have saved yourself all the labours you are now undertaking in the wreck of my fortunes.

8 So please go thoroughly into everything and let me have a full and reliable account. Wish me (as you do) to be *somebody*, since I can no longer be what I was or what I might have been. And please take me as reproaching *myself* in this letter, not you. I should be grateful if you would write letters and arrange for their dispatch to any persons you think ought to be written to in my name. Dispatched 17 August.

61 (III. 16)

Scr. Thessalonicae xii Kal. Sept. an. 58

CICERO ATTICO SAL.

Totum iter mihi incertum facit exspectatio litterarum vestrarum Kal. Sext. datarum. nam si spes erit, Epirum, si minus, Cyzicum aut aliud quid sequemur. tuae quidem litterae quo saepius a me leguntur hoc spem faciunt mihi minorem, quod cum ⟨diligentius⟩ lectae sunt tum id quod attulerunt ad spem infirmant, ut facile appareat te et consolationi servire et veritati. itaque te rogo plane ut ad me quae scies ut erunt, quae putabis ita scribas ut putabis. data XII Kal. [April.].

62 (III. 17)

Scr. Thessalonicae prid. Non. Sept. an. 58 (§3)

CICERO ATTICO SAL.

1 De Quinto fratre nuntii nobis tristes nec varii venerant ex a. d. IIII Non. Iun. usque ad prid. Kal. Sept. eo autem die Livineius, L. Reguli libertus, ad me a Regulo missus venit. is omnino mentionem nullam factam esse nuntiavit, sed fuisse tamen sermonem de C. Clodi filio; isque mihi tum a fratre litteras attulit. sed postridie Sesti pueri venerunt, qui a te litteras attulerunt non tam exploratas a timore quam sermo Livinei fuerat. sane sum in meo infinito maerore sollicitus, et eo magis quod Appi quaestio est.

Ep. 61] 2 nam si *s*: nam aliud aliquid si Ω.
4 quod Z^l: quam Ω.
5 diligentius *addidi*. lect(a)e *HNΔ*: lete *G*: laetae *Z*: lat(a)e *VR*.
7 itaque *Brunus* (*M*4): idque Ω. scies *Wesenberg*: scis Ω.
8 ut putabis Z^bλ: *om.* Ω. data Ω: D. λ. April. *del. Baiter*.

Ep. 62] 1, 2 a.d. IIII *scripsi*: ante diem Ω. non. iun(iis) usque δ: non unius que *al.* Ω. eo...attulit (*v.* 6) *om. GH*. eo *C*: ego Ω.

61 (III. 16)

Thessalonica, 19 August 58

CICERO TO ATTICUS

I am waiting for letters from you and others of date Kalends of August, and that makes my whole plan of travel uncertain. If there is hope I shall make for Epirus; if not, for Cyzicus or some place or other. The oftener I read your letters the less optimistic they make me. When carefully read they cast doubts on the reasons for optimism which they adduced. It's easy to see that you are trying to combine consolation with honesty. I do ask you therefore to write to me plainly, giving what you know just as it is, and what you think just as you think it. Dispatched 19th.

62 (III. 17)

Thessalonica, 4 September 58

CICERO TO ATTICUS

1 Gloomy and also consistent reports about my brother Quintus continued to reach me between 2 June and 29 August. On that day however, Livineius, L. Regulus' freedman, arrived, sent to me by Regulus. He told me that nothing had been said publicly, but there had been some talk of C. Clodius' son. He also brought me a letter from my brother. But the following day boys of Sestius' arrived bringing a letter from you which was not quite so reassuring as what Livineius had said. Naturally I feel anxiety, amid my own infinite distress, and all the more because Appius is President of the Court.

2–3 die Livineius *Victorius*: diei ei ve in eius *R*: diei ei (*om. s*) veniens *V*Δ: die (*spat. seq.*) eius *N*: die Livinus *C*: (die) Levineius λ.

3 L. *Manutius*: m *vel* in *V*Δ: *om. NRC.* libertus *vulg.*: l. Ω*C.*

5 tum a *V*: cum (quom) a *NOR*: $\widehat{\text{qm}}$ *M*: cum *bdm*: Quinto *s.*

8 livini *GRM.*

9 appi(i) qu(a)estio ΣδZ^{l}λ: -ispectio *M.*

2 Cetera quae ad me eisdem litteris scribis de nostra spe intellego esse languidiora quam alii ostendunt. ego autem, quoniam non longe ab eo tempore absumus in quo res diiudicabitur, aut ad te conferam me aut etiam nunc circum haec loca commorabor.

3 Scribit ad me frater omnia sua per te unum sustineri. quid te aut horter quod facis aut agam gratias, quod non exspectas? tantum velim fortuna det nobis potestatem ut incolumes amore nostro perfruamur. tuas litteras semper maxime exspecto; in quibus cave vereare ne aut diligentia tua mihi molesta aut veritas acerba sit. data prid. Non. Sept.

63 (III. 18)

Scr. Thessalonicae c. iv Id. Sept. an. 58

CICERO ATTICO SAL.

1 Exspectationem nobis non parvam attuleras cum scripseras Varronem tibi pro amicitia confirmasse causam nostram Pompeium certe suscepturum et, simul a Caesare ei litterae quas exspectaret remissae essent, actorem etiam daturum. utrum id nihil fuit an adversatae sunt Caesaris litterae an est aliquid in spe? etiam illud scripseras, eundem 'secundum comitia' dixisse.
2 fac, si vides quantis in malis iaceam et si putas esse humanitatis tuae, me fac de tota causa nostra certiorem. nam Quintus frater, homo mirus qui me tam valde amat, omnia mittit spei plena, metuens, credo, defectionem animi mei. tuae autem litterae sunt variae; neque enim me desperare vis nec temere sperare. fac, obsecro te, ut omnia quae perspici a te possunt sciamus.

3, 4 exopto *GN.*
5 cavere *R* vereare *Baiter*: -eriZ^1 ('*quamquam postrema syllaba non admodum appareat*'): *om.* Ω.
Ep. 63] **1**, 3 simul ac a *Mm.* ei *om. NMm.*
4 actorem *Z*: au(c)t- Ω.
2, 2 fac *om. EH.*

2 I perceive that the rest of your letter dealing with my own prospects strikes a less sanguine note than what I hear from other correspondents. However, as we are not far away from the time of decision one way or the other, I shall either move to your place or stay yet awhile around here.

3 My brother writes that you are in everything his sole support and stay. I won't encourage you to do what you are already doing or offer thanks which you do not expect. I can only wish that fortune may grant us to enjoy one another's affection in our native land. I always eagerly await your letters. Don't have any fear that I may find them tediously detailed or painfully candid. Dispatched 4 September.

63 (III. 18)

Thessalonica, c. 10 September 58

CICERO TO ATTICUS

1 My expectations were considerably aroused by what you told me about Varro's having assured you as a friend that Pompey would definitely take up my cause, and would further appoint an agent as soon as he had received a letter which he was expecting from Caesar. Did this mean nothing, or was Caesar's letter unfavourable, or is there still a hope? You also wrote that
2 he said 'after the elections'. Do, if you see the trouble I am in and think it becomes the kindness of your heart, do inform me how my whole case stands. My brother Quintus, a wonderful fellow to love me as he so ardently does, paints everything in rosy colours, fearing, I suppose, that I may otherwise give up altogether. Your letters vary. You don't want me to despair nor yet to hope unwisely. I beg you, let me know everything you can make out.

64 (III. 19)

Scr. Thessalonicae xvi Kal. Oct. an. 58 (§3)

CICERO ATTICO SAL.

1 Quoad eius modi mihi a vobis litterae adferebantur ut aliquid ex iis esset exspectandum, spe et cupiditate Thessalonicae retentus sum; postea quam omnis actio huius anni confecta nobis videbatur, in Asiam ire nolui, quod et celebritas mihi odio est et, si fieret aliquid a novis magistratibus, abesse longe nolebam. itaque in Epirum ad te statui me conferre, non quo mea interesset loci natura qui lucem omnino fugerem, sed et ad salutem libentissime ex tuo portu proficiscar et, si ea praecisa erit, nusquam facilius hanc miserrimam vitam vel sustentabo vel, quod multo est melius, abiecero. ⟨ero⟩ cum paucis; multitudinem dimittam.

2 Me tuae litterae numquam in tantam spem adduxerunt quantam aliorum; ac tamen mea spes etiam tenuior semper fuit quam tuae litterae. sed tamen, quoniam coeptum est agi, quoquo modo coeptum est et quacumque de causa, non deseram neque optimi atque unici fratris miseras ac luctuosas preces nec Sesti ceterorumque promissa nec spem aerumnosissimae mulieris Terentiae nec miserrimae [mulieris] Tulliolae obsecrationem et fidelis litteras tuas. mihi Epirus aut iter ad salutem dabit aut quod scripsi supra.

3 Te oro et obsecro, T. Pomponi, si me omnibus amplissimis, carissimis iucundissimisque rebus perfidia hominum spoliatum, si me a meis consiliariis proditum et proiectum vides, si

---

Ep. 64] 1, 1 quoad λ: quod Ω. litter(a)e a vobis Δ.
7–8 et ad λ: ad *EHR*: et *GM*: ut *NV*δ.
8 saltem *N*δ.
10 ero *add. vulg.*
2, 2 ac tamen *Wesenberg*: att- *GV*Δ: actum *N*: et t- *E*: et tum *R*.
3 tamen quoniam Ω: q- t- *m*: q- *c*. agi *Schütz*: age *E*: ago *GVRMm* (coeptum...modo *om. Nbd*).
7 mulieris *secl. Bentivolius*.

64 (III. 19)

Thessalonica, 15 September 58

CICERO TO ATTICUS

1 As long as my letters from you and my other friends were such as to arouse some expectation of result, hope and desire kept me at Thessalonica. When it appeared to me that all activity was over for the year, I did not feel like going to Asia because crowds are odious to me, and I did not want to be far away in case something might be done by the new magistrates. So I have decided to betake myself to your place in Epirus; not that the character of the locality makes any difference to me seeing that I shun the light of day altogether, but yours is the harbour from which I shall most gladly set out to my deliverance, or, if that is past praying for, there is nowhere else where I can more easily support this miserable existence, or, what is much better, discard it. I shall have few companions, I shall dismiss the throng.

2 Your letters have never made me so hopeful as other people's, and yet my hopes have always been even fainter than your letters. However, since a beginning has been made, no matter how or why, I shall not be found wanting. I owe that to my excellent and only brother's piteous, mournful entreaties, and the promises of Sestius and the rest, and the hopes of my unhappy wife, and the pleas of my poor little Tullia, and your loyal letters. Epirus will either set me on the road to safety or —what I said just now.

3 T. Pomponius, I beg and implore you: you see how the perfidy of others has stripped me of all that I enjoyed in such ample measure, all that I most valued and delighted in, how I have been betrayed and thrown to the wolves by my own

3, 1 T. (Tite) ΣZ: *om. HMm.*

intellegis me coactum ut ipse me et meos perderem, ut me tua misericordia iuves et Quintum fratrem, qui potest esse salvus, sustentes, Terentiam liberosque meos tueare, me, si putas te istic visurum, exspectes, si minus, invisas si potes, mihique ex agro tuo tantum adsignes quantum meo corpore occupari potest, et pueros ad me cum litteris quam primum et quam saepissime mittas. data XVI Kal. Oct.

65 (III. 20)

Scr. Thessalonicae iii Non. Oct. an. 58 (§4)

CICERO S. D. Q. CAECILIO Q. F. POMPONIANO ATTICO

1 Quod quidem ita esse et avunculum tuum functum esse officio vehementissime probo; gaudere me tum dicam si mihi hoc verbo licebit uti. me miserum! quam omnia essent ex sententia, si nobis animus, si consilium, si fides eorum quibus credidimus, non defuisset! quae colligere nolo ne augeam maerorem; sed tibi in mentem venire certo scio quae vita esset nostra, quae suavitas, quae dignitas. ad quae recuperanda, per fortunas incumbe, ut facis, diemque natalem reditus mei cura ut in tuis aedibus amoenissimis agam tecum et cum meis. ego huic spei et exspectationi quae nobis proponitur maxime tamen volui praestolari apud te in Epiro, sed ita ad me scribitur ut putem esse commodius nos eisdem in locis esse.

2 De domo et Curionis oratione, ut scribis ita est. in universa salute, si ea modo nobis restituetur, inerunt omnia: ex quibus nihil malo quam domum. sed tibi nihil mando nominatim, totum me tuo amori fideique commendo. quod te in tanta

---

3, 4 *prius* ut *EVRbds*: si *GNMm*.

Ep. 65] CICERO SALVTEM DICIT *Brunus* (M^4): CICERO ATTICO SALUTEM *VΔ*: *om. GNR*. Q. C(a)ecilio *GORΔC*: cecilio *N*: *om. V*. Q(u.) f. *GC*: Q.fr(atri) *N*(quinto) *VRΔ*. Pomponiano... sed (*v.* 6) *om. GH*. Pomponiano *O(?)PMC*: -oiano *R*: -onio *NVδ*.

advisers. You know how I was made to work my own and my family's ruin. Pity me then and help me. Support my brother Quintus, who can still be saved, look after Terentia and my children, wait for me if you think you are likely to see me in Italy, if not, visit me if you can and assign me so much of your land as will suffice to contain my body. And send me boys with letters as soon and as often as you can. Dispatched 15 September.

65 (III. 20)

Thessalonica, 5 October 58

CICERO TO Q. CAECILIUS Q. F. POMPONIANUS ATTICUS

1 Well, I heartily approve that it is so and that your uncle has done the proper thing. I shall say I am glad if the time ever comes when I can use that word. Ah me! How well content I should be in every way if courage, judgement, and the good faith of those I trusted had not failed me! I won't make a catalogue, which would add to my distress, but I am sure that the picture of my old life will come to your mind, its amenity, its dignity. Try, I beg you, to recover all this, as you are doing, and make it possible for me to celebrate the birthday of my return in your delightful house with you and my family. I *did* want to hold myself in readiness for this proffered hope and expectation on your property in Epirus for preference, but the letters I get suggest that it will be more convenient for me to stay where I am.

2 About my house and Curio's speech, it is just as you say. Everything will be included in the general restoration of my status, if only that comes about. I care as much about the house as anything. But I don't give you any specific charge, I throw myself entirely on your affection and loyalty. It is very good

---

1, 6 venire in mentem Δ. certo *GNΔc*: -te *HVR*.
9 amoenissimis *om. Mm*.

hereditate ab omni occupatione expedisti, valde mihi gratum est. quod facultates tuas ad meam salutem polliceris, ut omnibus rebus a te praeter ceteros iuver, id quantum sit praesidium video intellegoque te multas partis meae salutis et suscipere et posse sustinere neque ut ita facias rogandum esse.

3 Quod me vetas quicquam suspicari accidisse ad animum tuum quod secus a me erga te commissum aut praetermissum videretur, geram tibi morem et liberabor ista cura, tibi tamen eo plus debebo quo tua in me humanitas fuerit excelsior quam in te mea. velim quid videas, quid intellegas, quid agatur ad me scribas tuosque omnis ad nostram salutem adhortere. rogatio Sesti neque dignitatis satis habet nec cautionis. nam et nominatim ferri oportet et de bonis diligentius scribi, et id animadvertas velim. data III Non. Oct. Thessalonicae.

66 (III. 21)

Scr. Thessalonicae v Kal. Nov. an. 58

CICERO ATTICO SAL.

Triginta dies erant ipsi cum has dabam litteras per quos nullas a vobis acceperam. mihi autem erat in animo iam, ut antea ad te scripsi, ire in Epirum et ibi omnem casum potissimum exspectare. te oro ut si quid erit quod perspicias quamcumque in partem quam planissime ad me scribas et meo nomine, ut scribis, litteras quibus putabis opus esse ut des. data v Kal. Nov.

2, 8 te *om. Mm.*
3, 2 a *GRbms*: ad *NVMd.*
6 tuosque omneis *Cλ*: que omnis *RM*: qui o- *G*: omnisque *NVms*: qui omnis que *bd.*
8 ferri *c*: ferre Ω.
Ep. 66] 3 casum *P*: causam Ω.
5 planissime *VRMmC*: plen- *G.*

news that you have freed your hands from business although you have so large an inheritance to think of. I fully appreciate the value of your offer to put all your resources at the service of my restoration and your assurance that you will give me priority in all manner of assistance. I know that you are taking upon yourself a multiple share of the burden, that you can support it, and that you do not have to be asked to do it.

3 As you tell me not to imagine that any supposed offence or neglect on my part has made any impression upon your mind, I shall do as you wish and dismiss that worry from my thoughts; but none the less I shall feel all the more indebted to you for showing a more generous spirit towards me than I have shown to you. I hope you will write and tell me what you see and what you gather and what is going on, and encourage all your friends to work for my restoration. Sestius' bill is not satisfactory either from the standpoint of dignity or from that of security. The law should mention my name and be more carefully drafted with respect to my property, and I should be grateful if you would pay attention to these points. Dispatched 5 October, Thessalonica.

66 (III. 21)

Thessalonica, 28 October 58

CICERO TO ATTICUS

It is thirty whole days from the time of writing since I heard anything from you. I am now thinking of doing what I said earlier, going to Epirus, and making that the place where I shall await my fate, whatever it is to be. I beg you to write to me in the plainest possible terms of anything that comes within your purview, whatever way it tends, and to write in my name, as you promise, to such persons as you think needful. Dispatched 28 October.

67 (III. 22)

Scr. partim Thessalonicae c. xv Kal. Dec., partim Dyrrachi vi Kal. Dec. an. 58 (*§4*)

CICERO ATTICO SAL.

1 Etsi diligenter ad me Quintus frater et Piso quae essent acta scripserant, tamen vellem tua te occupatio non impedisset quo minus, ut consuesti, ad me quid ageretur et quid intellegeres perscriberes. me adhuc Plancius liberalitate sua retinet iam aliquotiens conatum ire in Epirum. spes homini est iniecta non eadem quae mihi, posse nos una decedere, quam rem sibi magno honori sperat fore. sed iam, cum adventare milites dicentur, faciendum nobis erit ut ab eo discedamus, quod cum faciemus ad te statim mittemus ut scias ubi simus.

2 Lentulus suo in nos officio, quod et re et promissis et litteris
declarat, spem nobis non nullam adfert Pompei voluntatis;
saepe enim tu ad me scripsisti eum totum esse in illius potestate.
de Metello scripsit ad me frater quantum speraret profectum
3 esse per te. mi Pomponi, pugna ut tecum et cum meis
mihi liceat vivere, et scribe ad me omnia. premor luctu
cum desiderio omnium rerum quae mihi me cariores semper
fuerunt. cura ut valeas.

4 Ego quod per Thessaliam si irem in Epirum perdiu nihil eram auditurus et quod [et] mei studiosos habeo Dyrrachinos ad eos perrexi, cum illa superiora Thessalonicae scripsissem. inde cum ad te me convertam, faciam ut scias, tuque ad me velim omnia quam diligentissime, cuicuimodi sunt, scribas. ego iam aut rem aut ne spem quidem exspecto. data VI Kal. Dec. Dyrrachi.

Ep. 67] **2,** 2 affert Σ: asserit *Mm*.
4 de *GNV*: et de *ORMm*.
3, 2 luctu *GNV*Δ: cum l- *ER*.
3 cum desiderio *s*: tum d- *ER*: d- cum *NMm*: d- *GV*. qu(a)e *ENVRbs*: qui *GMdm*. mihi me *VR*Δ: mihi *EGN*.
4, 2 mei *bds*: et mei Ω.
5 cuicuimodi ς: -do *Mdm*: cui modo *Rb*: quoquomodo *Gs*: quoiusmodi *V*.

67 (III. 22)

Dispatched from Dyrrachium, 25 November 58

CICERO TO ATTICUS

1 Though my brother Quintus and Piso have given me details of what has happened, I could have wished that pressure of business had not prevented you from sending your usual full accounts of what is going on and what you make of it. Plancius' generosity still keeps me where I am, though I have made several moves to get away to Epirus. The good soul is possessed by the hope, which I do not share, that we may be able to leave for home together, and thinks that will be a great feather in his cap. But I shall presently *have* to part from him, as soon as news comes of the soldiers arriving. When I do, I shall send you word at once, so that you know my whereabouts.

2 Lentulus' friendly attitude, evinced in actions, promises, and letters, makes me somewhat hopeful of Pompey's good will, as you have often told me that he is completely under Pompey's thumb. As to Metellus, my brother has written to me of how
3 much he hopes you have been able to achieve there. Strive, my dear fellow, to win me the right to live with you and my family, and write all to me. I am consumed by grief together with regret for all those things that have always been dearer to me than life itself. Keep well.

4 After writing the above at Thessalonica I proceeded to Dyrrachium, because if I had travelled to Epirus by way of Thessaly I should have heard nothing for a very long time, and because the townspeople are warm friends of mine. When I leave there for your place I shall let you know, and do you please write everything to me, good and bad, in the fullest detail. Results, not hopes, are what I expect now, results or nothing. Dispatched 25 November, Dyrrachium.

68 (III. 23)

Scr. Dyrrachi prid. Kal. Dec. an. 58 (§5)

CICERO ATTICO SAL.

1 A. d. v Kal. Dec. tris epistulas a te accepi, unam datam a. d. VIII Kal. Nov., in qua me hortaris ut forti animo mensem Ianuarium exspectem, eaque quae ad spem putas pertinere, de Lentuli studio, de Metelli voluntate, de tota Pompei ratione perscribis. in altera epistula praeter consuetudinem tuam diem non adscribis, sed satis significas tempus; lege enim ab octo tribunis pl. promulgata scribis te eas litteras eo ipso die dedisse, id est a. d. IIII Kal. Nov., et quid putes utilitatis eam promulgationem attulisse perscribis. in quo si iam nostra salus cum hac lege desperata erit, velim pro tuo in me amore hanc inanem meam diligentiam miserabilem potius quam ineptam putes; sin est aliquid spei, des operam ut maiore diligentia posthac a nostris magistratibus defendamur.

2 Nam ea veterum tribunorum pl. rogatio tria capita habuit; unum de reditu meo, scriptum incaute. nihil enim restituitur praeter civitatem et ordinem; quod mihi pro meo casu satis est, sed quae cavenda fuerint et quo modo te non fugit. alterum caput est tralaticium, de impunitate: 'si quid contra alias leges eius legis ergo factum sit'. tertium caput, mi Pomponi, quo consilio et a quo sit inculcatum vide.

Scis enim Clodium sanxisse ut vix aut omnino non posset nec per senatum nec per populum infirmari sua lex. sed vides numquam esse observatas sanctiones earum legum quae abrogarentur. nam si id esset, nulla fere abrogari posset; neque

---

Ep. 68] **1**, 1 decembr. *V*: septembr. *EGNR* Δ.
8 putes ς: potest Ω*C*: post *bds*.
9 perscribis Σ: sc- Δ. iam *GNR*: iam haec *VMm*: haec iam *bds*.
10 amore *Hs*: more *GN*Δ: m(o)erore *VR*.
2, 1 tribunorum...contentus (4, 16) *om. GH*.
8 aut omnino *ER*Δ: aut ut o- *NOP*: aut animo *V*.

68 (III. 23)

Dyrrachium, 29 November 58

CICERO TO ATTICUS

1 On 26 November I received three letters from you, one dispatched 25 October, in which you urge me to keep my courage up and wait for January; and you go through various items which you think have hopeful implications, Lentulus' zeal, Metellus' good will, Pompey's general policy. Your second letter, contrary to your custom, is undated, but you give sufficient indication as to when it was written, for you say you wrote on the very day that the eight Tribunes promulgated their law, i.e. 29 October, and you explain the good you think the promulgation has done. If by this time all hope of my restoration has gone the same way as the bill, I wish you may care enough for me to find my idle solicitude on the subject pitiable rather than foolish. If however there is still hope, I trust you will use your endeavours so that in future the magistrates who are on my side champion me with more circumspection.

2 That bill of the old Tribunes had three clauses, the first concerning my return. This was carelessly drafted, for it restores nothing except my citizenship and rank. That is enough for me in my present plight, but you must see what points ought to have been provided for and how. The second clause, on impunity, is common form: 'should there be any contravention of other laws in consequence of this law'. As for the third clause—my dear Pomponius, do ask yourself why it was put in and by whom.

Clodius, as you know, attached sanctions to his bill to make it almost or quite impossible for it to be invalidated either by the Senate or by the Assembly. But precedent shows you that such sanctions have never been observed in the case of laws to be repealed. Otherwise it would be virtually impossible to repeal

enim ulla est quae non ipsa se saepiat difficultate abrogationis. sed cum lex abrogatur, illud ipsum abrogatur quo non eam
3 abrogari oporteat. hoc cum et re vera ita sit et cum semper ita habitum observatumque sit, octo nostri tribuni pl. caput posuerunt hoc: 'si quid in hac rogatione scriptum est quod per leges plebisve scita' (hoc est, quod per legem Clodiam) 'promulgare, abrogare, derogare, obrogare sine fraude sua non liceat, non licuerit, quodve ei qui promulgavit, ⟨abrogavit⟩, derogavit, ⟨obrogavit⟩ ob eam rem poenae multaeve sit, E. H. L. N. R.'

4 Atque hoc in illis tribunis pl. non laedebat; lege enim collegi sui non tenebantur. quo maior est suspicio malitiae alicuius, cum id quod ad ipsos nihil pertinebat, erat autem contra me, scripserunt, ut novi tribuni pl., si essent timidiores, multo magis sibi eo capite utendum putarent. neque id a Clodio praetermissum est. dixit enim in contione a. d. III Non. Nov. hoc capite designatis tribunis pl. praescriptum esse quid liceret. tamen in lege nulla esse eius modi caput te non fallit; quod si opus esset, omnes in abrogando ⟨uterentur⟩. ut Ninnium aut ceteros fugerit, investiges velim, et quis attulerit, et qua re octo tribuni pl. ad senatum de me referre non dubitarint, sibi quod observandum illud caput non putabant, eidem in abrogando tam cauti fuerint ut id metuerent soluti cum essent quod ne iis

---

2, 13 abrogatur (*post* ipsum) $\Sigma Z^{(b)}\lambda$: -aretur Δ. quo non $Z^{(b)}$: quo modo Ω.
3, 1 quom et *VR*: et quod *N*: quod Δ. et cum (quom) Σ: *om.* Δ.
6 quodve ei *Ant. Augustinus*: quod vel Ω*c*.
6–7 abrogavit *et* obrogavit *add. Pithoeus.* E. H. L. N. R. *Manutius*: EH. INR *C*: eh mr *vel sim.* Ω. multaeve *v.l. ed. Hervagiana*: -ae ut Ω*C*.
4, 1 atque *etiam C*.
5 putarent Δ: -et Σ.
6 in *om.* Δ.
7 prescriptum *P*: persc- *NV*Δ: sc- *R*.
8 tamen *s*: tam Ω.
9 uterentur *add.* ς. ut *Malaespina*: aut Ω. Ninnium *vulg.* (*de Ant. F Malaespinae vix credendum*): nimium Ω.
11 dubitarint *Mm*: -rent *bds*: -runt *VR*: dubitant *N*. sibi *scripsi*: sive *vel* sine Ω.

any law, for there is none but protects itself by putting difficulties in the way of repeal. But when a law is repealed, the
3 provision against repeal is repealed at the same time. This being so, not only in fact but in traditional opinion and practice, our eight Tribunes put down the following clause: 'Should there be anything in the terms of this bill, whether in promulgation of a new statute or in repeal or restriction or contravention of an existing statute, which is or was forbidden by or punishable under the laws or the resolutions of the plebs' (which is to say under the lex Clodia) 'or which renders the person responsible for such promulgation or repeal or restriction or contravention liable to penalty or fine, nothing of this is enacted by the present bill.'

4 Now *those* Tribunes had nothing of the sort to fear, for they were not bound by a law emanating from one of their own body. That makes one all the more suspicious of some malicious intention when one finds them including a provision which did not affect themselves but was contrary to my interest, so that their successors, if timorously disposed, would feel under a much stronger compulsion to adopt the clause. Nor did that escape Clodius. At a meeting on 3 November he remarked that this clause laid down for the Tribunes-Designate how far they might go. And yet you must be aware that no law contains such a clause, though if there were any need for it everyone would have used it in repealing statutes. I should be grateful if you would inquire how it was that Ninnius and the others failed to see this, and who brought the clause in, and why the eight Tribunes, who did not hesitate to make a motion on my case in the Senate (because they did not regard themselves as bound by that clause), pushed caution so far when it came to the repeal that they were afraid of a sanction which did not

13 fuerint *Bosius*: -runt Ω (-rant *N*).

quidem qui lege tenentur est curandum. id caput sane nolim novos tribunos pl. ferre. sed perferant modo quidlibet; uno capite quo revocabor, modo res conficiatur, ero contentus. iam dudum pudet tam multa scribere; vereor enim ne re iam desperata legas, ut haec mea diligentia miserabilis tibi, aliis irridenda videatur. sed si est aliquid in spe, vide legem quam T. Fadio scripsit [T.] Visellius. ea mihi perplacet. nam Sesti nostri, quam tu tibi probari scribis, mihi non placet.

5 Tertia est epistula prid. Id. Nov. data, in qua exponis prudenter et diligenter quae sint quae rem distinere videantur, de Crasso, de Pompeio, de ceteris. qua re oro te ut, si qua spes erit posse studiis bonorum, auctoritate, multitudine comparata rem confici, des operam ut uno impetu perfringatur, in eam rem incumbas ceterosque excites. sin, ut ego perspicio cum tua coniectura tum etiam mea, spei nihil est, oro obtestorque te ut Quintum fratrem ames, quem ego miserum misere perdidi, neve quid eum patiare gravius consulere de se quam expediat sororis tuae filio, meum Ciceronem, cui nihil misello relinquo praeter invidiam et ignominiam nominis mei, tueare quoad poteris, Terentiam, unam omnium aerumnosissimam, sustentes tuis officiis. ego in Epirum proficiscar cum primorum dierum nuntios excepero. tu ad me velim proximis litteris ut se initia dederint perscribas. data prid. Kal. Dec.

4, 15 quidlibet λ: quid iubet (-ent *N*) Ω.
16 res conficiatur *s*: reco- Ω.
20 T. *del.* ς.
5, 9 quid eum *Em*: quidem *GNVΔ*: aliquid q- *R*.
11 tueare *Manutius*: tuere Ω.
14 tu Δ: tua Σ.

apply to them and which even those statutorily subject to it are at liberty to ignore. I sincerely hope that the new Tribunes will not propose this clause. However, let them only put *something* through, never mind what. I shall be satisfied with a single clause recalling me, provided only the thing be done. Already I feel ashamed to be writing at such length, and I fear that by the time you read this the matter may already have been given up as a bad job. In that case my solicitude will seem pitiable to you and ludicrous to others. But if there is anything left to hope for, take a look at the law which Visellius drafted for T. Fadius. It seems admirable to me. Sestius' draft, which you say you approve of, I don't like.

5 The third letter was dispatched on 12 November. In it you explain, knowledgeably and carefully, the factors which you think are holding matters back, covering Crassus, Pompey, and the rest. If therefore any hope emerges of carrying the business through by the support of the honest men, by influence, by a crowd brought up for the purpose, I beg you to try for a break-through at a single push. Put your heart into the effort and rouse the others. But if, as I see to be the case on your forecast and my own too, there is *no* hope, I beg and adjure you to care for my poor brother Quintus, whom I have ruined, unlucky wretch that I am. Don't let him take any decision concerning himself which might be to the injury of your nephew. Protect my Marcus as far as you can. Poor little boy, I am leaving him nothing but my hated and dishonoured name. Support Terentia with your good offices: no woman ever had so much to bear. I shall go to Epirus when I get news of the first few days. Please write to me in your next letter how the opening moves go. Dispatched 29 November.

69 (III. 24)

Scr. Dyrrachi iv Id. Dec. an. 58 (§2)

CICERO ATTICO SAL.

1 Antea, cum ad me scripsissetis vestro consensu consulum provincias ornatas esse, etsi verebar quorsum id casurum esset, tamen sperabam vos aliquid aliquando vidisse prudentius; postea quam mihi et dictum est et scriptum vehementer consilium vestrum reprehendi, sum graviter commotus, quod illa ipsa spes exigua quae erat videtur esse sublata. nam si tribuni pl. nobis suscensent, quae potest spes esse? ac videntur iure suscensere, cum et expertes consili fuerint ei qui causam nostram susceperant et nostra concessione omnem vim sui iuris amiserint, praesertim cum ita dicant, se nostra causa voluisse suam potestatem esse de consulibus ornandis, non ut eos impedirent sed ut ad nostram causam adiungerent; nunc, si consules a nobis alieniores esse velint, posse id libere facere; sin velint nostra causa, nihil posse se invitis. nam quod scribis, ni ita vobis placuisset, illos hoc idem per populum adsecuturos fuisse, invitis tribunis pl. fieri nullo modo potuit. ita vereor ne et studia tribunorum amiserimus et, si studia maneant, vinclum illud adiungendorum consulum amissum sit.

2 Accedit aliud non parvum incommodum quod gravis illa opinio, ut quidem ad nos perferebatur, senatum nihil decernere ante quam de nobis actum esset, amissa est, praesertim in ea

---

Ep. 69] **1,** 1 cum (quom) Σ: quam Δ.
5 sum graviter *GNΔλ*: g- s- *HVR*.
6 nam...amiserimus (2, 13) *om. GH.*
8 ei *Orelli*: et Ω.
9 vestra *P*.
10 cum ita *NPm*: ea ita *VRM*: cum ea ita *d*: cum ea *bs*.
13 velint (*alterum*) *Ernesti*: vellent Ω.
14 se *NΔ*: eis *VR*. ni δ: ne Ω.

69 (III. 24)

Dyrrachium, 10 December 58

CICERO TO ATTICUS

1 When you and my other supporters wrote to me that the grants for the consular provinces had been authorized with your consent, I had my misgivings about the consequences but I hoped that at some point or other you had seen further than I. Later I was told, both orally and in writing, that your tactics were coming in for strong criticism. I was deeply disturbed, because it looks as if even the faint ray of hope that existed has been extinguished. For if the Tribunes are annoyed with us, what hope can there be? And apparently they have fair reason for annoyance, since those of them who had undertaken to help our cause were not consulted and, through a concession on our part, have lost all the power of their prerogative—more especially as they say it was for our sake that they wanted the Consuls' grants to be under their control, not in order to make difficulties for them but to attach them to our cause. They point out that if the Consuls should now choose to take an unfriendly attitude towards us they are free to do so; whereas, should they be well disposed, they can do nothing against their (the Tribunes') wishes. As for your argument that if you had not agreed the Consuls would have carried their point through the Assembly, that would have been out of the question if the Tribunes had been in opposition. So I am afraid we may have lost the support of the Tribunes and, supposing that remains, I fear the hold we might have had over the Consuls has been let slip.

2 A further and not negligible disadvantage is that the firm impression (or so it was reported to me) that the Senate was not going to pass any decrees before my case had been dealt with has been let go by the board, and that for a purpose not

causa quae non modo necessaria non fuit sed etiam inusitata ac nova (neque enim umquam arbitror ornatas esse provincias designatorum), ut, cum in hoc illa constantia quae erat mea causa suscepta imminuta sit, nihil iam possit non decerni. iam iis ad quos relatum est amicis placuisse non mirum est; erat enim difficile reperire qui contra tanta commoda duorum consulum palam sententiam diceret. fuit omnino difficile non obsequi vel amicissimo homini Lentulo vel Metello, qui simultatem humanissime deponeret. sed vereor ne hos tamen tenere potuerimus, tribunos pl. amiserimus. haec res quem ad modum ceciderit et tota res quo loco sit velim ad me scribas et ita ut instituisti. nam ista veritas, etiam si iucunda non est, mihi tamen grata est. data IIII Id. Dec.

70 (III. 25)

Scr. Dyrrachi c. med. m. Dec. an. 58

CICERO ATTICO SAL.

Post tuum datae discessum litterae mihi Roma adlatae sunt ex quibus perspicio nobis in hac calamitate tabescendum esse. neque enim (sed bonam in partem accipies) si ulla spes salutis nostrae subesset, tu pro tuo amore in me hoc tempore discessisses. sed ne ingrati aut ne omnia velle nobiscum una interire videamur, haec omitto; illud abs te peto, des operam, id quod mihi adfirmasti, ut te ante Kal. Ian. ubicumque erimus sistas.

2, 7–8 iam iis (his) Σ: iis Δ.
10 fuit *N*Δ: et f- *VR*.
13 amiserimus *P*: mis- Ω.
Ep. 70] 1 datae *Watt*: a me Ω.
2 perspicio *H*: pros- *GVR*Δ: *om. N*.
3 accipies *Nms*: -peres *GVM*: -pe *R*. salutis ς: -ti Ω.
6 haec Σ: hoc Δ.

merely unnecessary but contrary to custom and precedent (I don't think provincial grants have ever been authorized for Designates). Thus there is nothing to stop any decree, now that the line determined in my interest has been breached in this instance. As for those friends of mine to whom the question was put, it is not surprising that they spoke in favour. It would have been difficult to find anyone prepared to raise his voice publicly against a proposal so much to the advantage of the two Consuls. No doubt it was difficult to refuse such a good friend as Lentulus, or Metellus either when he was so handsomely waiving his grudge against me. But I suspect that we could have held them all the same and that we have lost the Tribunes. I should be grateful if you would tell me how this came about and how the whole matter stands, and do so in your present manner. This frankness of yours may not make for pleasant reading, but I welcome it. 10 December.

70 (III. 25)

Dyrrachium, c. mid-December 58

CICERO TO ATTICUS

A letter from Rome, dispatched after you left, has reached me from which I see that I must pine away in this miserable state. And indeed (you won't take this amiss), if there were any lingering hope of my restoration, caring for me as you do you would not have left Rome at this time. But I leave this aside, for fear of appearing ungrateful or of seeming to want everything to go to ruin along with myself. I only ask you to try, as you assured me you would, to present yourself wherever I am before the Kalends of January.

71 (III. 26)

Scr. Dyrrachi, ut vid., med. m. Ian. an. 57

CICERO ATTICO SAL.

Litterae mihi a Quinto fratre cum senatus consulto quod de me est factum adlatae sunt. mihi in animo est legum lationem exspectare; et si obtrectabitur, utar auctoritate senatus et potius vita quam patria carebo. tu, quaeso, festina ad nos venire.

72 (III. 27)

Scr. Dyrrachi, ut vid., in. m. Febr. an. 57

CICERO ATTICO SAL.

Ex tuis litteris et ex re ipsa nos funditus perisse video. te oro ut quibus in rebus tui mei indigebunt nostris miseriis ne desis. ego te, ut scribis, cito videbo.

73 (IV. 1)

Scr. Romae c. iv Id. Sept. an. 57

CICERO ATTICO SAL.

1 Cum primum Romam veni fuit⟨que⟩ cui recte ad te litteras darem, nihil prius faciendum mihi putavi quam ut tibi absenti de reditu gratularer. cognoram enim, ut vere scribam, te in consiliis mihi dandis nec fortiorem nec prudentiorem quam me ipsum nec etiam propter meam in te observantiam nimium in custodia salutis meae diligentem; eundemque te, qui primis

---

Ep. 71] 4 *add. in fine epistulae* VI Cal. ianuar. *V*, data VI Kal. Ianuar. cura ut valeas *OR*.

Ep. 72] 1 et ex *VΔ*: ex *GNR*. 2 mei *Ps*: meis Ω. indigebunt *ms*: indulg- *G*: indulgebant *HNVRMbd*. 3 *post* videbo *add.* vale *H*, plura non scribo. cura ut valeas *VR*.

Ep. 73] 1 fuitque cui *s*: fuit cui *GNOP*: fuit qui *VMm*: que *R*.

71 (III. 26)

Dyrrachium (?), mid-January 57

CICERO TO ATTICUS

I have received a letter from my brother Quintus along with the decree about me passed in the Senate. I propose to wait for the laws to be put to the vote, and if there is opposition I shall avail myself of the Senate's authority and prefer loss of life to that of country. Pray make haste to join me.

72 (III. 27)

Dyrrachium (?), early February 57

CICERO TO ATTICUS

From your letter and from the facts themselves I see that I am utterly finished. In matters where my family needs your help I beg you not to fail us in our misery. According to your letter I shall see you soon.

73 (IV. 1)

Rome, about 10 September 57

CICERO TO ATTICUS

1 As soon as I arrived in Rome and came by a suitable person to take a letter to you, I considered it my first duty to congratulate you on my return. I had found you, to be quite frank, neither bolder nor wiser than myself as an adviser, nor, I may add, excessively sedulous in guarding me from harm in default of observances on my part to make you so. But I also found that, having in the early days shared my error, or rather

3 ut vere scribam *GNP*: ut resc- *VR*: te vere sc- (scibam *M*) *Mm*: ut bene sc- Z^l.

5 propter...observantiam ΩZ^l ('*plane scriptum*'): propter ita meam in te obs- Z^b.

temporibus erroris nostri aut potius furoris particeps et falsi
timoris socius fuisses, acerbissime discidium nostrum tulisse
plurimumque operae, studi, diligentiae, laboris ad conficiendum
2 reditum meum contulisse. itaque hoc tibi vere adfirmo, in
maxima laetitia et exoptatissima gratulatione unum ad cumulandum gaudium conspectum aut potius complexum mihi tuum defuisse. quem semel nactus si umquam dimisero ac nisi etiam praetermissos fructus tuae suavitatis praeteriti temporis omnis exegero, profecto hac restitutione fortunae me ipse non satis dignum iudicabo.

3 Nos adhuc in nostro statu quod difficillime recuperari posse arbitrati sumus, splendorem nostrum illum forensem et in senatu auctoritatem et apud viros bonos gratiam, magis quam opta⟨ra⟩mus consecuti sumus; in re autem familiari, quae quem ad modum fracta, dissipata, direpta sit non ignoras, valde laboramus, tuarumque non tam facultatum, quas nostras esse iudico, quam consiliorum ad colligendas et constituendas reliquias nostras indigemus.

4 Nunc, etsi omnia aut scripta esse a tuis arbitror aut etiam nuntiis ac rumore perlata, tamen ea scribam brevi quae te puto potissimum ex meis litteris velle cognoscere.

Prid. Non. Sext. Dyrrachio sum profectus, ipso illo die quo lex est lata de nobis. Brundisium veni Non. Sext. ibi mihi Tulliola mea fuit praesto natali suo ipso die, qui casu idem natalis erat et Brundisinae coloniae et tuae vicinae Salutis; quae res animadversa a multitudine summa Brundisinorum gratulatione

---

1, 7 potius *H. Stephanus*: totius Ω.
8 timoris ς: moris $Z^{(l)}$ ('*liber unus manuscr. longe optimus*'): rumoris Ω. discidium Σλ: dissi- *HMm*.

2, 4 quem semel...aestimabunt (7, 13) *desunt in GH*. si umquam *Meutzner*: ē umq- *R*: tum q- *M*: numq- *NV*δ.
5 fructus *VPms*: flu- *NOR* (-tos) *Mbd*. tuae *s*: tuos tu(a)e Ω.

3, 1 difficillime *ENVR*: -ille me *O*: -ile me Δ.
2 forensem *s*: -sium Ω.
3 bonos viros *c*.
4 optaramus *Manutius*: optamus Ω: -avimus *P*.

infatuation, and participated in my false alarm, you felt our severance most keenly, and devoted a vast amount of time and zeal and patience and labour to bringing about my return.

2 And so I sincerely assure you that in the plenitude of longed-for joy and congratulation one thing has been wanting to make my cup flow over: to see you, or rather to hold you in my arms. Once I win that happiness, if ever I let it go and if I do not also claim all the arrears of your delightful company that are owing to me, I shall really consider myself hardly worthy of this restitution of my fortunes.

3 Of my general position it can so far be said that I have attained what I thought would be most difficult to recover, namely my public prestige, my standing in the Senate, and my influence among the honest men, in larger measure than I had dreamed possible. But my private affairs are in a very poor way—you are aware how my property has been crippled and dissipated and pillaged—and I stand in need not so much of your resources, which I count as my own, as of your advice in pulling together what is left and putting it on a sound footing.

4 I shall now give you a brief account of such matters as I think you will particularly like to learn from my pen, though I expect you have already been informed of everything in letters from your own people or for that matter by report and general rumour.

I left Dyrrachium on 4 August, the very day on which the law for my recall was put to the vote. I landed at Brundisium on the Nones of August. My little Tullia was there to welcome me. It was her birthday and also, as it happened, the foundation day of the colony of Brundisium and of the temple of your neighbour the Goddess of Weal, a coincidence which attracted

4, 2 scribam *ENO*: insc- *VR*Δ.
8 a *om.* Δ.
8–9 gratulatione...est *C*: *om.* Ω.

celebrata est. a. d. †VI† Id. Sext. cognovi, cum Brundisi essem, litteris Quinti fratris mirifico studio omnium aetatum atque ordinum, incredibili concursu Italiae, legem comitiis centuriatis esse perlatam. inde a Brundisinis honestissime ornatus iter ita feci ut undique ad me cum gratulatione legati convenerint.

5 Ad urbem ita veni ut nemo ullius ordinis homo nomenclatori notus fuerit qui mihi obviam non venerit, praeter eos inimicos quibus id ipsum, se inimicos esse, non liceret aut dissimulare aut negare. cum venissem ad portam Capenam, gradus templorum ab infima plebe completi erant. a qua plausu maximo cum esset mihi gratulatio significata, similis et frequentia ⟨et⟩ plausus me usque ad Capitolium celebravit in foroque et in ipso Capitolio miranda multitudo fuit. postridie in senatu, qui fuit dies Non. Sept., senatui gratias egimus.

6 Eo biduo, cum esset annonae summa caritas et homines ad theatrum primo, deinde ad senatum concurrissent, impulsu Clodi mea opera frumenti inopiam esse clamarent, cum per eos dies senatus de annona haberetur et ad eius procurationem sermone non solum plebis verum etiam bonorum Pompeius vocaretur idque ipse cuperet multitudoque a me nominatim ut id decernerem postularet, feci et accurate sententiam dixi. cum abessent consulares, quod tuto se negarent posse sententiam dicere, praeter Messallam et Afranium, factum est senatus consultum in meam sententiam ut cum Pompeio ageretur ut eam rem susciperet, lexque ferretur. quo senatus consulto

---

4, 9 ante diem (ante *etiam C*) *Vms*: ante dies *NRMbd*. VI (sext.) *Vms*: sex *Md*: sextil. *b*: ë *R*: *om. N*. cum...essem *ΣZ*: *om.* Δ.
12 honestissime ς: -mis Ω.
13 fecimus *Mm*.

5, 1 ullius *ms*: il- Ω.
6 similis et *NV*Δ: -li *ER*.
7 et *s*: *om.* Ω.

6, 5 plebi λ.
7 feci et *P*δ: faciet Ω.
9 Afranium *Brunus* (*M^4*), *ut vid.*: aeranium *vel sim.* Ω. factum *bs*: -us Ω.

popular notice and was joyfully celebrated by the townsfolk. On 11 August (?), while at Brundisium, I learned by letter from Quintus that the law had been carried in the Assembly of Centuries amid remarkable demonstrations of enthusiasm by all ranks and ages and with an extraordinary concourse of country voters. Thence I set out, after receiving the most flattering marks of regard from the townspeople, and as I travelled along I was joined by congratulatory deputations from all quarters.

5 So I arrived at the outskirts of Rome. Not a man whose name was known to my nomenclator, no matter what his rank, but came out to meet me, except for enemies who could neither conceal nor deny the fact that they were such. When I reached the Porta Capena I found the steps of the temples thronged by the common people, who welcomed me with vociferous applause. Like numbers and applause followed me to the Capitol. In the Forum and on the Capitol itself the crowd was spectacular. In the Senate on the following day, the Nones of September, I made a speech of thanks to the House.

6 Two days later I spoke again. The price of grain had risen very high, and a crowd flocked first to the theatre and then to the Senate, clamouring at Clodius' instigation that the shortage was my doing. The Senate met during those days to consider the grain situation, and there was a general demand, not only from the populace but from the honest men too, that Pompey be asked to take charge of supplies. He himself was eager for the commission, and the crowd called on me by name to propose it. I did so in a full-dress speech. In the absence of all the Consulars except Messalla and Afranius, because, as they alleged, it was not safe for them to speak, the Senate passed a decree as proposed by me, to the effect that Pompey should be asked to undertake the matter and appropriate legislation be introduced. The decree was recited immediately, and the

recitato continuo, ⟨cum multitudo⟩ more hoc insulso et novo plausum meo nomine recitando dedisset, habui contionem. ⟨eam⟩ omnes magistratus praesentes praeter unum praetorem et duos tribunos pl. dederunt.

7 Postridie senatus frequens et omnes consulares. nihil Pompeio postulanti negarunt. ille legatos quindecim cum postularet, me principem nominavit et ad omnia me alterum se fore dixit. legem consules conscripserunt qua Pompeio per quinquennium omnis potestas rei frumentariae toto orbe terrarum daretur; alteram Messius, qui omnis pecuniae dat potestatem et adiungit classem et exercitum et maius imperium in provinciis quam sit eorum qui eas obtineant. illa nostra lex consularis nunc modesta videtur, haec Messi non ferenda. Pompeius illam velle se dicit, familiares hanc. consulares duce Favonio fremunt. nos tacemus, et eo magis quod de domo nostra nihil adhuc pontifices responderant. qui si sustulerint religionem, aream praeclaram habemus, superficiem consules ex senatus consulto aestimabunt; sin aliter, demolientur, suo nomine locabunt, rem totam aestimabunt.

8 Ita sunt res nostrae, 'ut in secundis fluxae, ut in adversis bonae'. in re familiari valde sumus, ut scis, perturbati. praeterea sunt quaedam domestica quae litteris non committo. Quintum fratrem insigni pietate, virtute, fide praeditum sic amo ut debeo. te exspecto et oro ut matures venire eoque animo venias ut me tuo consilio egere non sinas. alterius vitae quoddam initium ordimur. iam quidam qui nos absentis

---

6, 12 continuo *etiam C.* cum multitudo *addidi* (cum populus ς): *om. etiam C.* insulso *ΔC*: in sulto *N*: inconsulto *R*: in consilio *V*.

13 plausum meo *O*: -um eo *N*: -u meo *RΔ*.

14 eam *addidi* (quam *Wesenberg*).

7, 3 ad *EVR*: *om. NΔ*.

6 adiungit *c*: -nxit *Ω*.

11 responderunt *F*.

12 aream *bds*: arram *m*: aera ī *M* : (m)eram *NVR*. habebimus δ.

13 ex δ: es *N*: sex *VRM*. sin...aestimabunt *NORΔ*: *om. V*.

people applauded after the silly new-fangled fashion when my name was read out. I then addressed them at the invitation of all magistrates present except for one Praetor and two Tribunes.

7 The following day there was a large attendance in the House, including all the Consulars. Pompey was given everything he asked. In asking for fifteen Lieutenant-Commissioners he named me first, and said that I should be his *alter ego* for all purposes. The Consuls drafted a law giving Pompey control over grain supplies throughout the world for a period of five years. Messius proposed an alternative bill which gives him control over all moneys and in addition a fleet, an army, and authority in the provinces superior to that of their governors. Our consular law now looks quite modest; Messius' is felt to be intolerable. According to himself Pompey favours the former, according to his friends the latter. The Consulars are seething, Favonius at their head. I hold my tongue, a course to which I am the more inclined because the Pontiffs have not yet given any answer about my house. If they lift the religious sanction, I have a splendid site and the Consuls, under senatorial decree, will estimate the value of the building. If not, they will pull down the temple, let out a contract in their own name, and make an estimate for the whole.

8 So stand my affairs:

'"Unsettled", when our luck is in;
When out, we call it "fair".'

My financial position, as you know, is in very far from good order. Moreover, there are certain private matters which I don't trust to a letter. My brother is a paragon of affection, courage and loyalty, and I love him as I ought. I am looking forward to seeing you and beg you to make haste; and when you do come, come prepared to give me the benefit of your advice. It is a sort of second life I am beginning. Already, now

defenderunt incipiunt praesentibus occulte irasci, aperte invidere. vehementer te requirimus.

74 (IV. 2)

Scr. Romae in. m. Oct. an. 57

⟨CICERO ATTICO SAL.⟩

1 Si forte rarius tibi a me quam a ceteris litterae redduntur, peto a te ut id non modo neglegentiae meae sed ne occupationi quidem tribuas; quae etsi summa est, tamen nulla esse potest tanta ut interrumpat iter amoris nostri et offici mei. nam ut veni Romam, iterum nunc sum certior factus esse cui darem litteras; itaque has alteras dedi.

Prioribus tibi declaravi adventus noster qualis fuisset et quis
esset status atque omnes res nostrae quem ad modum essent,
2 'ut in secundis fluxae, ut in adversis bonae'. post illas datas
litteras secuta est summa contentio de domo. diximus apud
pontifices prid. Kal. Oct. acta res est accurate a nobis, et si
umquam in dicendo fuimus aliquid, aut etiam si numquam
alias fuimus, tum profecto dolor et ⟨rei⟩ magnitudo vim quan-
dam nobis dicendi dedit. itaque oratio iuventuti nostrae deberi
non potest; quam tibi, etiam si non desideras, tamen mittam
cito.

3 Cum pontifices decressent ita, 'si neque populi iussu neque plebis scitu is qui se dedicasse diceret nominatim ei rei praefectus esset neque populi iussu aut plebis scitu id facere iussus

---

Ep. 74] *novam ep. fecit Brunus* (M^4)*: continuant codd.*

1, 7 declaravi *NΔ*: -rat *H*: -ra *G*: -rabitur *VR*.

2, 4 numquam ς: um- Ω.

5 dolor et rei *Gulielmius*: dolor et *GNΔ*: -ris et *OR*: -ris *V*.

6 itaque...inquam (5, 12) *om. GH.* iuventuti...deberi *Victorius*: iubent uti (*sic etiam* Z^l)...debere *vel sim.* Ω.

3, 1 decressent *s*: -scent Ω. populi iussu ς: -li vis (-luus, -lus) Ω.

2 plebis scitu is CZ^b: plebiscitius *vel sim.* Ω. ei ς: et Ω.

that I am here, secret resentment and open jealousy are setting in among some of those who championed me when I was away. I need you badly.

74 (IV. 2)

Rome, beginning of October 57

CICERO TO ATTICUS

1 If so be that you hear from me less often than from others, let me ask you not to put it down to pressure of business, let alone negligence. True, I am extremely busy, but no amount of business can interfere with the course of our affection and the attention I owe you. The fact is, this is the second time since my return to Rome that I have heard of anyone to carry a letter; so this is my second letter.

In the first I gave you an account of the manner of my arrival, my present position, and the whole state of my affairs:

> '"Unsettled", when our luck is in;
> When out, we call it "fair".'

2 After I sent that letter there followed a tremendous struggle over my house. I addressed the Pontiffs on 29 September. I dealt faithfully with my theme, and if ever I amounted to anything as a speaker, or even if I never did at any other time, I think I can say that on that occasion intensity of feeling and the importance of the issue lent me a certain force of eloquence. So our younger generation cannot be kept waiting for the speech. I shall send it to you shortly, even if you are not anxious to have it!

3 The Pontiffs having found that 'that portion of the site might be restored to me without sacrilege, providing the person claiming to have consecrated it was not commissioned by name thereto by an order of the people or resolution of the

---

3 populi iussu aut *V*: -luus sua ut *vel sim.* *NRΔ*.

esset, videri posse sine religione eam partem areae mihi restitui', mihi facta statim est gratulatio (nemo enim dubita⟨ba⟩t quin domus nobis esset adiudicata), cum subito ille in contionem escendit quam Appius ei dedit. nuntiat [iam] populo pontifices secundum se decrevisse, me autem vi conari in possessionem venire; hortatur ut se et Appium sequantur et suam Libertatem ut defendant. hic cum etiam illi infimi partim admirarentur partim irriderent hominis amentiam (ego statueram illuc non accedere nisi cum consules ex senatus consulto porticum Catuli restituendam locassent), Kal. Oct.

4 habetur senatus frequens. adhibentur omnes pontifices qui erant senatores. a quibus Marcellinus, qui erat cupidissimus mei, sententiam primus rogatus quaesivit quid essent in decernendo secuti. tum M. Lucullus de omnium collegarum sententia respondit religionis iudices pontifices fuisse, legis ⟨es⟩se senatum; se et collegas suos de religione statuisse, in senatu de lege statuturos cum senatu. itaque suo quisque horum loco sententiam rogatus multa secundum causam nostram disputavit. cum ad Clodium ventum est, cupiit diem consumere, neque ei finis est factus; sed tamen, cum horas tris fere dixisset, odio et strepitu senatus coactus est aliquando perorare. cum fieret senatus consultum in sententiam Marcellini, omnibus praeter unum adsentientibus, Serranus ⟨intercessit. de⟩ intercessione statim ambo consules referre coeperunt. cum sententiae gravissimae dicerentur, senatui placere mihi domum

---

3, 4 eam *s*: iam Ω. areae mihi ς: ara emi (arae mi) *Mm*: aree emi *s*: aeraemi *R*: aeream *V*: aree *N*: *spat. bd.*
5 dubitabat *Corradus*: -tat Ω.
7 escendit *M*: asc- *N*δ: desc- *VR*. nuntiat ς: -ant Ω. iam *seclusi* (*habuit* Z^b).
9 possessionem *Vbds*: -ne *NRMm*.
10 ut *om. N.* infimi ς: infirmi Ω*C*.
4, 2 qui erat *om.* Σ.
4 tum M. *vulg.*: cum (quom) me Ω (cum M. *b*).

plebs, neither ordered so to act by an order of the people or resolution of the plebs', I was at once congratulated on the result, for nobody doubted that the house had been adjudged to me. Suddenly Clodius got up to address a meeting at Appius' invitation. He announced to the people that the Pontiffs had found in his favour, but that I was attempting to get possession by force, and urged them to follow Appius and himself and defend their Liberty. Even those groundlings were astonished, some of them, while others laughed at his folly—I had decided not to go near the place until the Consuls by decree of the Senate had contracted for the rebuilding of Catulus' portico. On the Kalends of October there was a meeting of the Senate,
4 well attended. All the Pontiffs who were Senators were called in. Marcellinus, who was very strongly on my side, as the first called upon, asked them to give reasons for their decree. M. Lucullus, speaking for all his colleagues, then replied that the Pontiffs had been judges of the religious issue, but the Senate was judge of the law. His colleagues and himself had given their verdict on the former; on the latter they would decide in the Senate, as Senators. Accordingly all of them, as called upon in their turn, spoke at length in favour of my case. When Clodius' turn came he set himself to talk out the sitting and there was no putting a stop to him. However, after holding forth for something like three hours, he was finally forced to wind up by noisy interruptions from the exasperated House. A decree in accordance with Marcellinus' proposal, unanimous apart from one dissentient, was in process of passage when Serranus interposed his veto. Both Consuls immediately set about referring the veto to the Senate. Some very powerful speeches were delivered to the effect that it was the Senate's

---

5 legis esse *Seyffert*: legisse *VR*: legis *NΔ*.
7 statuturos ς: statuos *NOΔ*: -ues *VR*: stat. *C*. cum...suo *C*: *om*. Ω.
13 intercessit. de *add. Victorius*.

restitui, porticum Catuli locari, auctoritatem ordinis ab omnibus magistratibus defendi, si quae vis esset facta, senatum existimaturum eius opera factam esse qui senatus consulto intercessisset, Serranus pertimuit et Cornicinus ad suam veterem fabulam rediit; abiecta toga se ad generi pedes abiecit. ille noctem sibi postulavit. non concedebant; reminiscebantur enim Kal. Ian. vix tamen ei de mea voluntate concessum est.

5 Postridie senatus consultum factum est id quod ad te misi. deinde consules porticum Catuli restituendam locarunt; illam porticum redemptores statim sunt demoliti libentissimis omnibus. nobis superficiem aedium consules de consili sententia aestimarunt HS vicies, cetera valde illiberaliter: Tusculanam villam quingentis milibus, Formianum HS ducentis quinquaginta milibus. quae aestimatio non modo vehementer ab optimo quoque sed etiam a plebe reprehenditur. dices 'quid igitur causae fuit?' dicunt illi quidem pudorem meum, quod neque negarim neque vehementius postularim. sed non est id; nam hoc quidem etiam profuisset. verum iidem, mi T. Pomponi, iidem, inquam, illi, quos ne tu quidem ignoras, qui mihi pinnas inciderant, nolunt easdem renasci. sed, ut spero, iam renascuntur. tu modo ad nos veni; quod vereor ne tardius interventu Varronis tui nostrique facias.

6 Quoniam acta quae sint habes, de reliqua nostra cogitatione cognosce. ego me a Pompeio legari ita sum passus ut nulla re impedirer. quod nisi vellem mihi esset integrum ut, si comitia

---

4, 16 locari ς: -re *N*Δ: iocare *VR*.
18 factam *scripsi*: -tum Ω.
22 ei *scripsi*: tibi Ωλ: sibi δ.
5, 1 senatus consultum *m*: -us consulto Ω: s. c. *bds*.
5 vicies *s*λ: vicis *vel sim.* Ω. illiberaliter *vulg.*: in te lib- Ω.
11 mi *VP*: mei *NOR*Δ.
6, 3 ut *Hofmann*: aut Ω.

desire that my house should be restored, a contract for Catulus' portico put in hand, and the authority of the House defended by all magistrates; further, that in case of any violence the person vetoing the decree would be held responsible by the Senate. Serranus took fright, and Cornicinus repeated his old charade—flinging off his toga and throwing himself on his knees before his son-in-law. Serranus asked for a night's grace, but the House, remembering the Kalends of January, was unwilling to allow it. However, with much ado and with my good will, the point was conceded him.

5 On the following day the decree which I send you herewith was passed. The Consuls then signed a contract for the restoration of Catulus' portico, and the contractors immediately demolished the other portico much to everybody's satisfaction. The Consuls with their assessors valued my house, that is the building, at HS 2,000,000 and the other properties at very ungenerous figures—the Tusculan villa at 500,000, the Formian at HS 250,000. The valuation is sharply criticized not only by better-class people but by the populace as well. You may wonder why this happened. *They* say my modesty was the reason, in that I neither refused compensation nor pressed my claim with vigour. But it isn't that. *That* would have been in my favour rather than otherwise. No, my dear T. Pomponius, those same gentry (you don't need me to tell you their names) who formerly clipped my wings don't want to see them grow back to their old size. However, I hope they *are* growing already. Only do come back and join me. I am afraid you may do so less promptly than you would have done, now that your (and my) friend Varro's visit has intervened.

6 Since you now know what has taken place, let me tell you of my plans for the future. I have let Pompey nominate me one of his Lieutenant-Commissioners, but on the understanding that my hands are left quite free. If I did not want to leave it

censorum proximi consules haberent, petere possem, votivam legationem sumpsissem prope omnium fanorum, lucorum; sic enim nostrae rationes utilitatis meae postulabant. sed volui meam potestatem esse vel petendi vel ineunte aestate exeundi, et interea me esse in oculis civium de me optime meritorum non alienum putavi.

7 Ac forensium quidem rerum haec nostra consilia sunt, domesticarum autem valde impedita. domus aedificatur, scis quo sumptu, qua molestia. reficitur Formianum, quod ego nec relinquere possum nec videre. Tusculanum proscripsi, ⟨etsi⟩ suburbano non facile careo. amicorum benignitas exhausta est in ea re quae nihil habuit praeter dedecus; quod sensisti tu absens, ⟨tui⟩ praesentes. quorum studiis ego et copiis, si esset per meos defensores licitum, facile essem omnia consecutus; quo in genere nunc vehementer laboratur. cetera quae me sollicitant μυστικώτερα sunt. amamur a fratre et a filia. ⟨te⟩ exspectamus.

75 (IV. 3)

Scr. Romae viii Kal. Dec. an. 57 (§5)

CICERO ATTICO SAL.

1 Avere te certo scio cum scire quid hic agatur tum ea a me scire, non quo certiora sint ea quae in oculis omnium geruntur si a

---

6, 4 possem ς: -ent Ω (-int *N*): posse $Z^{(b)}$.
5 sumpsissem *C*: -isse *NRΔ*: suppressisse *V*. lucorum *Bosius*: loc- Ω.
6 utilitatis *H*: -tes ΩZ^{l}.
9 alienum *Gs*: -norum *HNVRΔC*.
7, 3 reficitur *Manutius*: -ciatur Ωλ.
4 etsi *add. Kayser.*
5 non *om. Mm.*
7 tui praesentes *Tyrrell–Purser*: praesentes *GVRΔ*: -ens es *H*: -ens praesentes *N*.
9 nunc *P*: hunc *R*: hinc *GVΔ*: hic *N*.
10 te *add. Manutius.*
11 ex(s)pectamus *GRMbd*: -ur *HNVPms*.

open to myself to stand for the Censorship if next year's Consuls hold an election to that office, I should have taken a votive commission to practically 'every temple and sacred wood'. That is what our calculations of my interest required. But I wanted to give myself the option of either standing or leaving at the beginning of next summer, and in the meanwhile I thought it would not be a bad idea to show myself to my fellow-countrymen, to whom I owe so much.

7 Such are my plans as to public affairs. As to my private life, they are in a terrible tangle. My house is being built, you know at what expense and trouble. My Formian villa is being reconstructed, and I cannot bear to let it go nor yet to look at it. I have put up the Tusculan property for sale, though I can't easily do without a place near Rome. The generosity of my friends was exhausted in a matter which yielded nothing but disrepute, as you saw in absence and your people saw on the spot. Relying on their loyalty and resources, if only my defenders had permitted it, I should have had ample for every purpose; whereas now I am greatly embarrassed in this respect. My other anxieties are more *sub rosa*. I have the affection of my brother and daughter. I look forward to seeing you.

75 (IV. 3)

Rome, 23 November 57

CICERO TO ATTICUS

1 I am sure you are dying to know what's afoot here, and also to know it from me—not that news of what goes on in full public view is any more reliable from my pen than when it

Ep. 75] 1, 1 avere...nuntiantur (*v.* 3) *om. GH.* avere te *EVbs*: avete *NR*: avere *OMdm.* certo *NOΔ*: -te *VR*: *om. E.* agatur *O(?)RΔ*: agitur *NV.* ea Ω: mea *PZ*[l].
2 sint *Vms*: sunt *ENORMbd.*

me scribuntur quam cum ab aliis aut scribuntur tibi aut nuntiantur, sed velim perspicias ex meis litteris quo animo ea feram quae geruntur et qui sit hoc tempore aut mentis meae sensus aut omnino vitae status.

2 Armatis hominibus a. d. III Non. Nov. expulsi sunt fabri de area nostra, disturbata porticus Catuli, quae ex senatus consulto consulum locatione reficiebatur et ad tectum paene pervenerat, Quinti fratris domus primo fracta coniectu lapidum ex area nostra, deinde inflammata iussu Clodi inspectante urbe coniectis ignibus, magna querela et gemitu non dicam bonorum, qui nescio an nulli sint, sed plane hominum omnium. ille ve⟨l ante⟩ demens ruere, post hunc vero furorem nihil nisi caedem inimicorum cogitare, vicatim ambire, servis aperte spem libertatis ostendere. etenim antea, cum iudicium nolebat, habebat ille quidem difficilem manifestamque causam, sed tamen causam; poterat infitiari, poterat in alios derivare, poterat etiam aliquid iure factum defendere: post has ruinas, incendia, rapinas desertus a suis vix iam †Decimum† dissignatorem, vix Gellium retinet, servorum consiliis utitur; videt, si omnis quos vult palam occiderit, nihilo suam causam difficiliorem quam adhuc sit in iudicio futuram.

3 Itaque a. d. III Id. Nov. cum Sacra via descenderem, insecutus est me cum suis. clamor, lapides, fustes, gladii; et haec improvisa omnia. discessi in vestibulum Tetti Damionis. qui

---

1, 3 scribuntur *Wesenberg*: -bantur Ω. scribuntur *V*: -bantur *ENORΔ*.
4 sed velim *EVR*: velim (*incipiente ep.*) *G*: sed ut v- *N*: sed ut verum Δ.
2, 7 nulli *Lambinus*: ulli Ω. ille...futuram (*v.* 17) *om. GH.*
7–8 vel ante demens *scripsi* (demens *iam Pius*): vehemens Ω.
8 ruere *etiam* *CZ*^{l}λ. vero *om.* Δ. nisi c(a)edem Σ*s*: sic edem *vel sim.* Δ.
12 poterat (*post* causam) *ER*: id p- *NVΔ*.
13 et poterat aliquid etiam *VR*.
14 dissignatorem *scripsi*: desi- Ω.

comes to you from the letters or reports of others, but I should like you to see from a letter of my own how I react to developments, and my attitude of mind and general state of being at the present time.

2 On 3 November an armed gang drove the workmen from my site, threw down Catulus' portico which was in process of restoration by consular contract under a senatorial decree and had nearly reached the roof stage, smashed up my brother's house by throwing stones from my site, and then set it on fire. This was by Clodius' orders, with all Rome looking on as the firebrands were thrown, amid loud protest and lamentation—I won't say from honest men, for I doubt whether they exist, but from all and sundry. Clodius was running riot even before, but after this frenzy he thinks of nothing but massacring his enemies, and goes from street to street openly offering the slaves their freedom. Earlier on, when he would not stand trial, he had a difficult, obviously bad case, but still a case. He could have denied the charges or blamed others or even have defended this or that action as legitimate. But after this orgy of wrecking, arson, and loot, his followers have left him. It is all he can do to keep Decius (?) the undertaker or Gellius, and he takes slaves for his advisers. He sees that if he slaughters everybody he chooses in broad daylight, his case, when it comes to court, won't be a jot worse than it is already.

3 Accordingly, on 11 November as I was going down the Via Sacra, he came after me with his men. Uproar! Stones flying, cudgels and swords in evidence. And all like a bolt from the blue! I retired into Tettius Damio's forecourt, and my

---

16 occiderit *ERbs*: -rint *OMdm*: -runt *NV*.

3, 2 et *om*. Δ.

3 discessimus Δ. vestibulum Tettii *Victorius*: -lo M. tetii *vel sim*. *NVRMm*: -lo *EG* (Tetti...futura (4, 3) *om*. *GH*).

erant mecum facile operas aditu prohibuerunt. ipse occidi potuit; sed ego diaeta curare incipio, chirurgiae taedet. ille omnium vocibus cum se non ad iudicium sed ad supplicium praesens trudi videret, omnis Catilinas Acidinos postea reddidit. nam Milonis domum, eam quae ⟨est in⟩ Cermalo, prid. Id. Nov. expugnare et incendere ita conatus est ut palam hora quinta cum scutis homines eductis gladiis, alios cum accensis facibus eduxerit. ipse domum P. Sullae pro castris sibi ad eam impugnationem sumpserat. tum ex Anniana Milonis domo Q. Flaccus eduxit viros acris; occidit homines ex omni latrocinio Clodiano notissimos, ipsum cupivit, sed ille †ex interiorem aedium Sulla se in†. senatus postridie Id. domi Clodius. egregius Marcellinus, omnes acres. Metellus calumnia dicendi tempus exemit adiuvante Appio, etiam hercule familiari tuo, de cuius constantia vitae tuae verissimae litterae. Sestius furere. ille postea, si comitia sua non fierent, urbi minari. ⟨Milo⟩ proposita Marcellini sententia, quam ille de scripto ita dixerat ut totam nostram causam areae, incendiorum, periculi mei iudicio complecteretur eaque omnia comitiis anteferret, proscripsit se per omnis dies comitialis de caelo servaturum.
4 contiones turbulentae Metelli, temerariae Appi, furiosissimae Publi; haec tamen summa, nisi Milo in campo obnuntiasset, comitia futura.

A. d. XII Kal. Dec. Milo ante mediam noctem cum manu

3, 5 diaeta (*sic etiam s*) curare *Cobet*: dia et accurari Ω (dia *om. NV*, *spat. rel.*). 7 trudi ς: videri Ω (praesens v- *om. VP*). Acidinos *vulg.*: sac- Ω. 8 eam...Cermalo *Turnebus*: meamque ceramio Ω. 10 eductis ς: ded- Ω. 13 acris *s*: agris Ω. 14 notissimos ς: -mo Ω. interiorem *R*Δ: -re *NVP*. 15 se *om. V*. senatus *Orelli*: -tu Ω: -tum *c*. 17 Appio *M corr.*: op(p)io Ω. 18 vitae tuae Σ: virtute Δ. 19 minari δ: mirari Ω. Milo *add. Wesenberg* (*post* proscripsit *iam Klotz*). 23 se per ς: super Ω*C*.
4, 2 mil(l)o in (-oī) *NR*: miloni *VP*Δ. campo Σ: -pum Δ.

companions had no difficulty in keeping out the rowdies. Clodius himself could have been killed, but I am becoming a dietician, I'm sick of surgery. When he found that everyone was calling for him to be bundled off to trial or rather to summary execution, his subsequent behaviour was such as made every Catiline look like an Acidinus. On 12 November he tried to storm and burn Milo's house in the Cermalus, bringing out fellows with drawn swords and shields and others with lighted firebrands, all in full view at eleven o'clock in the morning. He himself had made P. Sulla's house his assault base. Then out came Q. Flaccus with some stout warriors from Milo's other house, the Anniana, and killed off the most notorious bandits of the whole Clodian gang. He had every wish to kill their principal, but *he* had gone to earth in the recesses (?) of Sulla's house. Senate on the 14th. Clodius at home. Marcellinus first-rate, and the rest backed him up vigorously. But Metellus talked out the time with a filibuster, abetted by Appius and also, I must add, by that friend of yours, about whose consistency of conduct you write most truly. Sestius was beside himself. Clodius later threatened reprisals against the city if his elections were not held. Milo on his side posted up Marcellinus' proposal, which the latter had read out from script, calling for a trial to cover my whole case—the site, the fires, and my own narrow escape, all to take place before the elections. He also put up an announcement that he would
4 watch the skies on all comitial days. Public speeches followed, a seditious one by Metellus, a reckless one by Appius, a quite frantic one by Publius. What it all came to was that, unless Milo declared contrary auspices in the Campus Martius, the elections would be held.

On 19 November Milo went to the Campus before

4 ante mediam noctem *HNVRbdsC*: media nocte *Mm*: *om.* *G*.
4–5 manu magna *NVRMmλ*: magna manu *G*.

magna in campum venit. Clodius, cum haberet fugitivorum delectas copias, in campum ire non est ausus. Milo permansit ad meridiem mirifica hominum laetitia, summa cum gloria. contentio fratrum trium turpis, fracta vis, contemptus furor. Metellus tamen postulat ut sibi postero die in foro obnuntietur; nihil esse quod in campum nocte veniretur; se hora prima in comitio fore. itaque a. d. XI Kal. [Ian.] in comitium Milo de nocte venit. Metellus cum prima luce furtim in campum itineribus †prope† deviis currebat; adsequitur inter lucos hominem Milo, obnuntiat. ille se recipit magno et turpi Q. Flacci convicio. a. d. X Kal. nundinae. contio biduo nulla.

5 A. d. VIIII Kal. haec ego scribebam hora noctis nona. Milo campum iam tenebat. Marcellus candidatus ita stertebat ut ego vicinus audirem. Clodi vestibulum vacuum sane mihi nuntiabatur: pauci pannosi sine lanterna. meo consilio omnia illi fieri querebantur, ignari quantum in illo heroë esset animi, quantum etiam consili. miranda virtus est. nova quaedam divina mitto, sed haec summa est: comitia fore non arbitror, reum Publium, nisi ante occisus erit, fore a Milone puto, si se in turba ei iam obtulerit occisum iri ab ipso Milone video. non dubitat facere, prae se fert; casum illum nostrum non extimescit. numquam enim cuiusquam invidi et perfidi consilio est us⟨ur⟩us nec inerti nobili⟨tati⟩ crediturus.

4, 6 delectas ς: deie- Ω.
8 contemptus δ: -entus ΩZ^{l}.
11 XII *Mm.* Ian. *del. Corradus.*
13 adsequitur...crediturus (5, 12) *om. GH.*
14 recipit *NVP*: recepit *OR*Δ.

5, 1 VIIII *R*λ: XIIII *NV*: VIII *O*Δ. scribebam *O*δ: scribam *NRM*(?): scripsi *V.*
3 nuntiabatur *N*: -bantur *VR*Δ.
4 pauci pannosi *sF*: -cis -si *VRMdm*: -cis -sis *Nb.* sine lanterna *s*: sine alanterna(m) *vel sim. VR*Δ: mea lanternam *N.* meo consilio *Manutius*: eo conscio Ω (-ntio *R*).
5 heroe ς: hero Ω.
6 mitto ς: mittite Ω.
8 se in turba ei *Klotz, Baitero duce*: se uti turb(a)e iam *NVCZ* (*sed* ut it urbae Z^{b}): sentitur hanc (ac *P*) iam *R*: -tur ve(h)iam Δ.
11 usurus *Klotz*: usus Ω.

midnight with a large following. Though Clodius had a picked force of runaway slaves at his back, he did not dare go to the Campus. Milo stayed till noon, to the public's enormous glee and his own great *réclame*. The campaign of the three brethren became a fiasco. They found their violence outmatched and their fury treated with contempt. However, Metellus asked Milo to declare the auspices to him next day in the Forum. No need, he said, to go at night to the Campus. He himself would be in the Comitium at daybreak. So, on the 20th, in came Milo to the Comitium while it was still dark. As dawn broke, there was Metellus scurrying along the byways to the Campus. Milo caught him up between the Woods and made his declaration, on which Metellus turned tail, to the accompaniment of a deal of coarse jeering from Q. Flaccus. The 21st was market-day, and for a couple of days there was no assembly.

5 On the night of the 22nd I am writing this between two and three o'clock. Milo is already in position on the Campus. My neighbour Marcellus (the candidate) is snoring loud enough for me to hear him. I am told that Clodius' forecourt is pretty well deserted—a handful of ragamuffins without a lantern. Clodius' party complain that it's all been my plan. Little do they know our heroic Milo, what a resourceful as well as gallant fellow he is. His spirit is amazing. I pass over certain recent brilliancies, but the sum and substance is as follows: I don't believe there will be any elections. I think Publius will be brought to trial by Milo, unless he is killed first. If he now puts himself in Milo's way in a rough-and-tumble I don't doubt that Milo will dispatch him with his own hands. He has no qualms about doing so, and makes no bones about it. He is not scared of what happened to me, for *he* is never going to follow anybody's envious and treacherous advice or put his trust in a sluggish nobility.

12 nobilitati *scripsi*: nobili Ω.

6 Nos animo dumtaxat vigemus, etiam magis quam cum florebamus; re familiari comminuti sumus. Quinti fratris tamen liberalitati pro facultatibus nostris, ne omnino exhaustus essem, illo recusante subsidiis amicorum respondemus. quid consili de omni nostro statu capiamus te absente nescimus. qua re adpropera.

76 (IV. 4)

Scr. Romae iii Kal. Febr. an. 56

CICERO ATTICO SAL.

Periucundus mihi Cincius fuit a. d. III Kal. Febr. ante lucem; dixit enim mihi te esse in Italia seseque ad te pueros mittere. quos sine meis litteris ire nolui, non quo haberem quod tibi, praesertim iam prope praesenti, scriberem, sed ut hoc ipsum significarem, mihi tuum adventum suavissimum exspectatissimumque esse. qua re advola ad nos eo animo ut nos ames, te amari scias. cetera coram agemus. haec properantes scripsimus. quo die venies, utique cum tuis apud me sis.

77 (IV. 7)

Scr. in Arpinati c. Id. Apr. an. 56

CICERO ATTICO SAL.

1 Nihil εὐκαιρότερον epistula tua, quae me sollicitum de Quinto nostro, puero optimo, valde levavit. venerat horis duabus ante Chaerippus, mera monstra nuntiarat.

De Apollonio quod scribis, qui illi di irati, homini Graeco qui conturbat atque idem putat sibi licere quod equitibus

6, 1 quam ς': quod *N*: *om. VRΔ*.
3 liberalitati ς': -atis Ω.
6 *in fine ep.* cura ut valeas *add. VR*.
Ep. 76] 5 expect- *P*: spect- Ω.
Ep. 77] **1**, 1 *verba* nihil εὐκαιρότερον *huic ep. tribuit* Z^l, 83 (IV. 6) Ω.
3 mera monstra *R*: -am -am *GNOΔ* (*de cod. V vide Vol. 1, p. 78*).
5 conturbat atque idem *Bosius*: -baret quidem Ω.

6 My *heart* is high, higher even than in my palmy days, but my purse is low. None the less, with the help of my friends and against his opposition, I am repaying my brother's generosity, so far as my resources allow, so as not to be left entirely penniless. With you away I don't know what line to take as to my position in general. So make haste.

76 (IV. 4)

Rome, 28 January 56

CICERO TO ATTICUS

Cincius is a very welcome arrival (before daybreak, 28 January), for he tells me that you are in Italy and that he is sending you boys. I didn't want them to go without a letter from me, not that I have anything to write to you about, especially as you are almost here, but just to tell you that I *am* more than delighted to hear of your return and have been impatiently looking forward to it. So hurry to join me, confident in my affection as in your own. I write in haste. The day you get here, don't fail to stay with me and bring your folks.

77 (IV. 7)

Arpinum, c. 13 April 56

CICERO TO ATTICUS

1 Nothing could have been more *à propos* than your letter, which has greatly relieved my anxiety about our nephew, good boy that he is. Chaerippus had arrived two hours previously with the most horrifying reports.

As for what you say about Apollonius, the gods confound him! The impudence of a Greek going bankrupt and thinking himself entitled to the same privilege as Roman Knights!

quod...cautior (2, 6) *om. GH.*

2 Romanis! nam Terentius suo iure. de Metello, ‘οὐχ ὁσίη φθιμένοισιν’, sed tamen multis annis civis nemo erat mortuus cui equidem. . . . tibi nummi meo periculo sint. quid enim vereris? quemcumque heredem fecit, nisi Publium fecit, †verum fecit non improbi quemquam† fuit ipse. quare in hoc thecam nummariam non retexeris, in aliis eris cautior.

3 Mea mandata de domo curabis, praesidia locabis, Milonem admonebis. Arpinatium fremitus est incredibilis de Laterio. quid quaeris? equidem dolui. ‘ὁ δὲ οὐκ ἐμπάζετο μύθων.’

Quid superest? etiam. puerum Ciceronem curabis et amabis, ut facis.

78 (IV. 4*a*)

Scr. Anti c. xi Kal. Quint., ut vid., an. 56

⟨CICERO ATTICO SAL.⟩

1 Perbelle feceris si ad nos veneris. offendes dissignationem Tyrannionis mirificam librorum meorum, quorum reliquiae multo meliores sunt quam putaram. et velim mihi mittas de tuis librariolis duos aliquos quibus Tyrannio utatur glutinatoribus, ad cetera administris, iisque imperes ut sumant membranulam ex qua indices fiant, quos vos Graeci, ut opinor,
2 σιττύβας appellatis. sed haec, si tibi erit commodum. ipse vero utique fac venias, si potes in his locis adhaerescere, et Piliam adducas; ita enim et aequum est et cupit Tullia.

2, 2–3 cui equidem *scripsi*: quid quidem Ω.
4 quemcumque *NOδ*: quaec- *M*: quem tum *R*.
5 improbi *NORbd*: -be *Mms*. quamquam *s*.
3, 4 quid *Manutius*: quod Ω.
5 amabis ut facis *Ps*: -it ut facias (-at *N*) Ω.
Ep. 78] *novam ep. fecit Brunus* (*M^4*), *ut vid.; ep.* 76 (IV. 4) *adnectunt codd.*
1, 1 veneris *GNΔ*: venies *VR*. dissign- *scripsi*: design- Ω.
2 bibliotheca (-am *N*) *post* meorum *add. NΔ*.
3 et Σ: etiam Δ. velim *Ps*: vellem Ω.

2 Terentius after all is within his rights. As for Metellus, *de mortuis*. All the same, no countryman of ours has died this many years past, of whom for my part.... Your money is safe enough, I'll guarantee. What are you worrying about? Whoever he's made his heir (unless it be Publius) can't be a greater rascal than he was himself (?), so you won't have to loosen your purse strings on his account, and you'll be more careful next time!

3 Please remember what I asked you to do about my house, post the guards and say a word to Milo. The people of Arpinum are in an amazing fume about Laterium. Well, there it is! *I* was very sorry, but 'little he recked my rede'.

No more, except to ask you to look after the boy and keep a soft spot in your heart for him, as you do.

78 (IV. 4*a*)

Antium, c. 20 June (?) 56

CICERO TO ATTICUS

1 It will be delightful of you to pay us a visit. You will find that Tyrannio has made a wonderful job of arranging my books. What is left of them is much better than I had expected. And I should be grateful if you would send me a couple of your library clerks to help Tyrannio with the gluing and other operations, and tell them to bring a bit of parchment for the
2 labels, σιττύβαι as I believe you Greeks call them. But this is if convenient to you. As for yourself, do come without fail, if you can bear to stick down here for a while, and bring Pilia, as is right and proper and as Tullia much wishes.

---

5 iisque *m*: is que *RM*: hisque *GNVP*.
7 σιττύβας *Tyrrell*: sillabos Ω.
2, 3 adducas *scripsi* (adduc *Orelli*): -cere Ω.

Me dius fidius, ne tu emisti ludum praeclarum! gladiatores audio pugnare magnifice. si locare voluisses, duobus his muneribus liberasses. sed haec posterius. tu fac venias, et de librariis, si me amas, diligenter.

79 (IV. 8)

Scr. Anti paulo post superiorem

CICERO ATTICO SAL.

1 Multa me in epistula tua delectarunt, sed nihil magis quam patina tyrotarichi. nam de rauduscolo quod scribis, 'μήπω μέγ' εἴπῃς πρὶν τελευτήσαντ' ἴδῃς'.

Aedificati tibi in agris nihil reperio. in oppido est quiddam de quo est dubium sitne venale, ac proximum quidem nostris aedibus. hoc scito, Antium Buthrotum esse Romae ut Corcyrae illud tuum. nihil quietius, nihil alsius, nihil amoenius. †ειμη-
2 ισητω† φίλος ⟨οἶ⟩κος. postea vero quam Tyrannio mihi libros disposuit, mens addita videtur meis aedibus. qua quidem in re mirifica opera Dionysi et Menophili tui fuit. nihil venustius quam illa tua pegmata, postquam sittybae libros illustrarunt. vale.

Et scribas mihi velim de gladiatoribus, sed ita bene si rem gerunt; non quaero male si se gesserunt.

2, 4 ne tu emisti *G*: nec veīsti *R*: nec (et *N*) veni- *NVP*: ne tu veni- *Mm*. ludum *Ernesti*: locum Ω.
5 audio *GN*Δ: enim a- *VR*.

Ep. 79] *hac ep. carent GH*.
1, 2 de rauduscolo *Victorius*: perau(i)d- *vel sim.* Ω*C*.
7 alsius *vulg.*: sa- Ω.
8 φίλος οἶκος *Victorius*: ΦΙΔΟϹ.ΚΟϹ *vel sim. RMm*.

Well, upon my word you have bought a fine outfit! I hear the gladiators are fighting magnificently. If you had cared to hire them out you would have cleared your outlay in these last two shows. But of this later. Do come, and *please* don't forget about the clerks.

79 (IV. 8)

Antium, shortly after 78 (IV. 4a)

CICERO TO ATTICUS

1 Many things in your letter delighted me, but the plate of fish *au gratin* was best of all. As for what you say about the bit of brass, 'boast not before thou seest the end of all'.

I can't find anything for you in the way of a house in the country. There is something in the town, which may or may be for sale, and very near our house too. Let me tell you that Antium is the Buthrotum of Rome as your Buthrotum is of Corcyra—the quietest, coolest, pleasantest place in the
2 world. 'Dwelling ever dear to me (?).' And now that Tyrannio has put my books straight, my house seems to have woken to life. Your Dionysius and Menophilus have worked wonders over that. Those shelves of yours are the last word in elegance, now that the labels have brightened up the volumes. Good-bye—oh, and you might let me know about the gladiators, but only if they give a good account of themselves. Otherwise the less I hear of them the better.

2, 2 mens *NC*: -sa *R*Δ.
4 sittybae *scripsi*: sit tibae Z^l: sit tibi *NR Ant.*: misit tibi *O*: misit Δ.
5 vale *Gronovius*: valde ΩZ^b.
7 si se gesserunt *Wesenberg* (si se gessere *Malaespina*): sit (sic *Ant. F*) egisse re *OR Ant. F*: egisse sic re(perias: *cf. ep.* 82 (IV. 8a) *init.*) *N*: sic egisse Δ.

80 (IV. 5)

Scr. Anti paulo post superiorem

CICERO ATTICO SAL.

1 Ain tu? an me existimas ab ullo malle mea legi probarique quam ⟨a⟩ te? cur igitur cuiquam misi prius? urgebar ab eo ad quem misi et non habebam exempla duo. quin etiam (⟨iam⟩ dudum enim circumrodo quod devorandum est) subturpicula mihi videbatur esse παλινῳδία. sed valeant recta, vera, honesta consilia. non est credibile quae sit perfidia in istis principibus, ut volunt esse et ut essent si quicquam haberent fidei. senseram, noram inductus, relictus, proiectus ab iis. tamen hoc eram animo ut cum iis in re publica consentirem. iidem erant qui fuerant. vix aliquando te auctore resipivi.

2 Dices ea te[nuisse] suasisse quae facerem, non etiam ut scriberem. ego mehercule mihi necessitatem volui imponere huius novae coniunctionis ne qua mihi lice⟨re⟩t ⟨re⟩labi ad illos qui etiam tum cum misereri mei debent non desinunt invidere. sed tamen modici fuimus ἀποθεώσει, ut scrips⟨eram⟩. erimus uberiores, si et ille libenter accipiet et hi subringentur qui villam me moleste ferunt habere quae Catuli fuerat, a Vettio me emisse non cogitant, qui domum negant oportuisse me

---

Ep. 80] 1, 1 ain...παλινῳδία (*v.* 5) *om. GH.* ain *O Ant. FΔZ*^{l}: an *NR*: an me *NΔZ*: me *OR*: me me *Ant. F.* mea legi *O(?)P*: mea (me a) lege *Nbds*: me ad lege (adl-) *RMm.*

2 a *Ps*: *om.* Ω.

3 exempla duo *Constans*: -plari *RM*: -plare *Od*: -plaria *N*: -plar *Pbms.* quin *Rinkes*: quid Ω. iam *add. Watt.*

8 iis *O*: is *R*: his *GNPΔ.* eram *C*: erat in Ω.

9 iis *NOδ*: his *G*: usi *RM.*

10–2, 1 resipivi (*sic etiam* λ) dices ς: res ipsi (ipi) iud- Ω: resipui iud- *s.*

2, 1 ea te *scripsi*: ea tenuisse Ω. facerem *Pius*: feceram Ω.

2 volui *vulg.*: nolui Ω.

3 ne qua mihi liceret *Victorius*: neque mihi (me *Mm*) licet Ω. relabi *Pluygers*: labi Ω.

5 sed tamen...oportere (*v.* 9) *om. GH.* ἀποθεώσει ς: ΑΠΟΘωCΙ *RMm.* scripseram. erimus *scripsi*: -serimus *NRMbd*: -seramus *P*: -simus *ms.*

80 (IV. 5)

CICERO TO ATTICUS

Antium, soon after 79 (IV. 8)

1 Come now! Do you really think there is anyone whom I would sooner have read and approve my compositions than yourself? Why then did I send this one to anybody else first? Because the person to whom I sent it was pressing me and I did not have two copies. There was also the fact (I might as well stop nibbling at what has to be swallowed) that I was not exactly proud of my palinode. But good night to principle, sincerity, and honour! You will scarcely credit the treachery of our public leaders, as they set up to be and *would* be if they had a grain of honesty about them. I had seen it, knew it, led on by them as I was, deserted, thrown to the wolves. Yet even so I was disposed to agree with them in politics. They proved to be what they had always been. At long last, and by your advice, I have come to my senses.

2 You will say that you recommended what I should *do*, not that I should write as well. The truth is, I wanted to bind myself irrevocably to this new alliance so as to make it quite impossible for me to slip back to those people who won't give up their jealousy even when they ought to be sorry for me. However, I have observed moderation in my 'apotheosis', as I told you I should. I shall give myself more rein if *he* receives the offering cordially and if, on the other hand, it wrings the withers of certain gentlemen who object to my owning a villa which once belonged to Catulus without recollecting that I bought it from Vettius, and say in the same breath that I ought not to have built my house and that I ought to sell it. But what's

---

6 hi *NR* (*sed* his ubr-) Δ: ii *C*. subringentur *vulg.*: -igentur Ω*C*.
7 a Vettio me Z^b: a vectio *O*(?)*R* *Ant.* *F*: avercio *N*: aut etiam Δ.

aedificare, vendere aiunt oportere. sed quid ad hoc si, quibus sententiis dixi quod et⟨iam⟩ ipsi probarent, laetati sunt tamen me contra Pompei voluntatem dixisse? finis sit. quoniam qui nihil possunt ii me nolunt amare, demus operam ut ab iis qui
3 possunt diligamur. dices 'vellem iam pridem'. scio te voluisse et me asinum germanum fuisse. sed iam tempus est me ipsum a me amari, quando ab illis nullo modo possum.

4 Domum meam quod crebro invisis est mihi valde gratum. viaticam Crassipes praeripit. tu 'de via recta in hortos?' videtur commodius. ad te postridie scilicet; quid enim tua? sed viderimus.

Bibliothecam mihi tui pinxerunt cum structione et sittybis. eos velim laudes.

81 (IV. 12)

Scr. Anti ex. m. Iun., ut vid., an. 56

CICERO ATTICO SAL.

Egnatius Romae est, sed ego cum eo de re †Halimeti† vehementer Anti egi. graviter se acturum cum Aquilio confirmavit. videbis ergo hominem si voles. Macroni vix videor praesto esse. Idibus enim auctionem Larini video et biduum praeterea. id tu, quoniam Macronem tanti facis, ignoscas mihi velim. sed si me diligis, postridie Kal. cena apud me cum P[et]ilia. prorsus id facies. Kalendis cogito in hortis Crassipedis quasi in

---

2, 9 oportere *scripsi*: -tuisse Ω. si *Pius*: sed Ω.
10 etiam *scripsi*: et Ω.
11 sit *Graevius*: sed Ωλ.

3, 2 me...sed *Corradus*: me ast num germanum fuisset? *C*: me ast num geram manum fuisset *NOΔ et sim. R*: me (*ceteris omissis*) *G*.

4, 2 viaticam...laudes (*v*. 6) *om. G*. viaticam *scripsi*: -cum Ω.
5 cum structione *Birt*: const- Ω. sit tybis *Mm*: sic tibi *R*: si scribis *Nbds*.

Ep. 81] 1 halimeti *GNΔC*: alimati *O*: alimenti *R*: -eti *P*.
2 anti(i) Ωλ: ante δ.

that compared with the fact that when I made speeches in the Senate on lines which even they approved they were delighted none the less that I had spoken against Pompey's wishes? *Il faut en finir.* Since the powerless won't be my friends, let me
3 try to make myself liked by the powerful. You will say that I might have thought of that sooner. I know you wanted me to do so, and that I have been a prize donkey. But now it's time for me to love myself since *they* won't love me whatever I do.

4 Thank you very much for keeping a close eye on my house. Crassipes is forestalling your welcome-home dinner. 'Straight from the road to the suburbs?' Well, it seems more convenient. I'll come to you the next day, naturally. It can't make any odds to you. But we shall see.

Your people have painted my library together with the book-cases and labels. Please commend them.

81 (IV. 12)

Antium, end of June (?), 56

CICERO TO ATTICUS

Egnatius is in Rome, but I took up Halimetus' (?) affair with him strongly at Antium. He assured me that he would speak seriously to Aquilius. So you will see him if you wish. I hardly think I can oblige Macro. There is the auction, you see, at Larinum on the Ides and for the two days following. As you think such a lot of Macro, I must ask you to forgive me. But don't fail to dine with me on the 2nd with Pilia. You really must. On the Kalends I propose to dine at Crassipes' place in

---

4 Idibus Σ (*sed* graviter...biduum *om. GH*) *Z*: *om.* Δ. biduum *N*: vid- *ORZ*^{l}: id- Δ.
5 mihi *GNP*: mi *OR codd. Mal.* Δ: me *H*.
6 Pilia ς: petilia Ω*C*.

deversorio cenare; facio fraudem senatus consulto. inde domum cenatus, ut sim mane praesto Miloni. ibi te igitur videbo; et praemonebo. domus te nostra tota salutat.

82 (IV. 8*a*)

Scr. Anti vel. in Tusculano c. xiv Kal. Dec., ut vid., an. 56

⟨CICERO ATTICO SAL.⟩

1 Apellas vix discesserat, cum epistula. quid ais? putasne fore ut legem non ferat? dic, oro te, clarius; vix enim mihi exaudisse videor. verum statim fac ut sciam, si modo tibi est commodum. ludis quidem quoniam dies est additus, eo etiam melius hic eum diem cum Dionysio conteremus.

2 De Trebonio, prorsus tibi adsentior. de Domitio,

'σύκῳ, μὰ τὴν Δήμητρα, σῦκον οὐδὲ ἓν
οὕτως ὅμοιον γέγονεν'

quam est ista περίστασις nostrae, vel quod ab iisdem vel quod praeter opinionem vel quod viri boni nusquam; unum dissimile, quod huic merito. nam de ipso casu nescio an illud melius. quid enim hoc miserius quam eum qui tot annos quot habet designatus consul fuerit fieri consulem non posse, praesertim cum aut solus aut certe non plus quam cum altero petat? si vero id est, quod nescio an sit, ut non minus longas iam in codicillorum fastis futurorum consulum paginulas habeant quam factorum, quid illo miserius nisi res publica, in qua ne speratur quidem melius quicquam?

3 De Natta ex tuis primum scivi litteris. oderam hominem.

---

10 praemonebo *Malaespina*: prom- Ω (promov- *R*) *Z*.

Ep. 82] *novam esse ep. vidit Manutius; cum* 79 (IV. 8) *iungunt codd.*

1, 1 Apellas...exorsus (3, 3) *om. G.* Apellas *Turnebus*: aperias *ORΔ* (*de N vide ad 79. 3, 7*): apenas *Ant. F.* quid ς: quod Ω.

3 tibi ΣZ^b: *om.* Δ.

the suburbs in lieu of an inn, and thus cheat the decree! Then home after dinner so that I can be ready for Milo in the morning. I shall see you there then, and shall warn you beforehand. All my household wish to be remembered to you.

82 (IV. 8*a*)

Antium or Tusculum, c. 17 Nov. (?) 56

CICERO TO ATTICUS

1 Hardly had Apellas left when your letter arrived. Are you really saying that you don't think he'll bring the bill to a vote? Please speak a little louder, I don't think I quite caught! But do let me know at once, if you conveniently can that is. As the games are going on for an extra day, I shall be all the happier to be spending that day here with Dionysius.

2 I quite agree about Trebonius. As for Domitius, 'No pea, by Jove, was e'er so like to pea' as this crisis of his to my own. The same people responsible, the same unexpectedness, the same conspicuous absence of the honest men. The only difference is that he asked for it. As for the actual blow, I dare say I came off the better of the two. Could anything be more lamentable than for a man who has been Consul-Designate from his cradle to be debarred from becoming Consul, especially as he has no competitors, or at any rate no more than one? And if it is true, as I dare say it is, that in the lists in their little notebooks the future Consuls take up as many pages as the past ones, could any condition be more pitiable than his—except the state's, in which no improvement is so much as hoped for?

3 Your letter gave me first news about Natta. I detested the

2, 4 nostrae *Manutius*: -ra Ω.
11 habeant *Gronovius*: -eat Ω.
3, 1 scivi *s*: -vit Ω.

de poëmate quod quaeris, quid si cupiat effugere? quid? sinam? de Fabio Lusco quod eram exorsus, homo peramans nostri semper fuit nec mihi umquam odio. satis enim acutus et permodestus ac bonae frugi. eum, quia non videbam, abesse putabam: audivi ex Gavio hoc Firmano Romae esse hominem et fuisse adsiduum. percussit animum. dices 'tantulane causa?' permulta ad me detulerat non dubia de Firmanis fratribus. quid sit quod se a me remov⟨er⟩it, si modo removit, ignoro.

4 De eo quod me mones, ut et πολιτικῶς me geram et τὴν ἔσω γραμμὴν teneam, ita faciam. sed opus est maiore prudentia, quam a te, ut soleo, petam. tu velim ex Fabio, si quem habes aditum, odorere et istum convivam tuum degustes et ad me de his rebus et de omnibus cottidie scribas. ubi nihil erit quod scribas, id ipsum scribito. cura ut valeas.

83 (iv. 6)

Scr. in Cumano c. xiv Kal. Apr., an. 55

CICERO ATTICO SAL.

1 De Lentulo scilicet sic fero ut debeo. virum bonum et magnum hominem et in summa magnitudine animi multa humanitate temperatum perdidimus, nosque malo solacio sed non nullo tamen consolamur quod ipsius vicem minime dolemus, non ut Saufeius et vestri, sed mehercule quia sic amabat patriam ut mihi aliquo deorum beneficio videatur ex eius incendio ereptus.

3, 2 sinam *scripsi*: -at Ω.
3–4 nostri semper Σ*c*: s- n- Δ.
9 removerit *Wesenberg*: removit Ω (*sed* si modo removit *om. GN*).
4, 1 πολιτικῶς λ: pol(l)iticos Ω.
1–2 ἔσω γραμμὴν *Corradus*: ΕΩΤΡΑΜΗΝ *al. GRMm*.
3 ex *Baiter*: *om.* Ω.
6 id ς: et Ω.
Ep. 83] 1, 3 nosque...viderimus (4, 2) *paucis exceptis om. G.* nullo *Malaespina*: multo Ω.
5 vestri sed *Malaespina*: -ris *EOR*Δ: nostre *N*.

man. You ask about the poem. Suppose it were to wish to spread its wings, what then? Should I let it? As for Fabius Luscus, as I had started to say, he was always extremely friendly to me and I never had anything against him—a clever fellow enough, very modest and worthy. Well, as I saw nothing of him, I thought he must be out of town, but I heard from Gavius of Firmum (you know) that he was in Rome and had been all the time. That gave me a jolt. You may think it a small thing to trouble about, but he had brought me a lot of reliable information about the brethren of Firmum. Why he should have withdrawn himself from me, if he really did, I have no idea.

4 As for your admonition to behave like a *politique* and to keep to the inner row, I shall do so. But I need greater *expertise*, and that, as usual, I shall ask you to supply. I should be glad if you would pump Fabius, if you have any opportunity, and sample that dinner-companion of yours, and write to me about this and all else every day. When you have nothing to say, why, say just that! Take care of your health.

83 (IV. 6)

Cumae, about 19 April 55

CICERO TO ATTICUS

1 The news about Lentulus affects me as it should, naturally. We have lost a good and great man, who combined a really lofty spirit with much grace and kindliness of manner. My consolation, poor enough but still a consolation, is that I feel no sorrow at all on *his* account—I don't say so after the fashion of Saufeius and your co-sectaries, but because upon my soul it seems to me a gift of providence that a man who loved his country as he did should be snatched away from its conflagration.

---

6 aliquo deorum (de e- *R*) *ER*: aliquid d- *O*: aliquid e- *N*Δ. ereptus *EOR*: esse e- *N*Δ.

nam quid foedius nostra vita, praecipue mea? nam tu quidem, etsi es natura πολιτικός, tamen nullam habes propriam
2 servitutem, communi †fueris non ne†. ego vero, qui, si loquor de re publica quod oportet, insanus, si quod opus est servus existimor, si taceo, oppressus et captus, quo dolore esse debeo? quo sum scilicet, hoc etiam acriore quod ⟨ne⟩ dolere quidem possum ut non in te ingratus videar. quid si cessare libeat et in oti portum confugere? nequiquam, immo etiam in bellum et in castra. ergo erimus ὀπαδοὶ qui ταγοὶ esse noluimus? sic faciendum est; tibi enim ipsi, cui utinam semper paruissem, sic video placere. reliquum est 'Σπάρταν ἔλαχες, ταύταν κόσμει'. non mehercule possum et Philoxeno ignosco qui reduci in carcerem maluit. verum tamen id ipsum mecum in his locis commentor ut ista [im]probem, idque tu cum una erimus confirmabis. a te litteras crebro ad me scribi video, sed omnis uno tempore accepi. quae res etiam auxit dolorem meum. casu enim trinas ante legeram quibus meliuscule Lentulo esse scriptum erat. ecce quartae fulmen! sed ille, ut scripsi, non miser, nos vero ferrei.

3 Quod me admones ut scribam illa Hortensiana, in alia incidi, non immemor istius mandati tui, sed mehercule incipiendo refugi ne qui videor stulte illius amici [non] intemperiem tulisse rursus stulte iniuriam illius faciam illustrem si quid scripsero, et

---

1, 9 servitutem *C*: virtutem Ω. communi (-is *N*) fueris (fru- *s*) non ne (nonne) *ORΔλ*: *om. E.*

2, 4 ne *Brunus* (M^4): *om.* Ω.

5 in te ingratus *Nbds*: integ- (in te g-) *EGRMm*. libeat *RM*: -et *EGNOP*: debet *m*.

6 nequiquam *Baiter*: ne(c)quicquam *EGNPΔ*: nequaq- *R*.

7 ὀπαδοὶ qui ταγοὶ *Victorius*: ΟΠΛΟΙ qui ΤΑΤΟΙ *vel sim. RMm*. noluimus *OMC*: vo- *Rm*: volemus *N*: nolumus (vol-) *Pbds*.

9 reliquum est *Corradus*: -quia est $\Delta Z^{(b)}$: -qui est λ: -qui a se *N*: -quia *vel* -qua (*sic s*) est *codd. Mal.*: quia est *OR*.

12 probem *Ant.*: imp- $\Omega Z^{(b)}$.

15 meliuscule ς: -us qu(a)e *GNΔ*: -us *ORs*.

17 ferrei *s*: ferri Ω: ferei *m*.

After all what could be more ignominious than the life we lead, I especially? For you, though you are a political animal by nature, are not subject to any peculiar servitude,
2 you * * * in common with the rest. But as for me, reckoned a madman if I speak on politics as I ought, a slave if I say what is expedient, and a helpless captive if I say nothing—how am I to feel? As I do I suppose, and all the more bitterly because I can't even grieve without seeming ungrateful to you. Suppose I choose to fold my hands and seek a haven of refuge in retirement? Vain thought! On the contrary I must join the fray. Am I then to be a camp follower having refused to be a general? Needs must, for I see that you yourself (if only I had always listened to you!) think it the best course. There is nothing else left. 'Sparta is your portion: embellish it'—no, confound it, I *can't*! I don't blame Philoxenus for preferring to be sent back to gaol. And yet I am trying down here to discipline myself to do just that, to approve their proceedings, and you will strengthen my resolution when we meet. I see you are writing to me often, but I received all your letters in one batch. That actually cost me an extra pang. I happened to open first the three in which you wrote that Lentulus was a trifle better. Then the thunderstroke in the fourth! But, as I have said, he is not to be pitied; *we* have hides of leather.

3 You remind me about writing the Hortensius piece. I have dropped into other things, but it was not that I had forgotten your injunction. But upon my word, I no sooner got started than I ran back again. I felt that I already looked silly enough for putting up with his unreasonable behaviour when he was my friend, without committing the further folly of drawing attention to his misconduct towards me by writing anything.

3, 3 qui videor *Lamb.* (*marg.*): quod video Ω. non *seclusi.*
4 illius ς: -is Ω. illustrem *P*: -re Ω.

simul ne βαθύτης mea, quae in agendo apparuit, in scribendo
4 sit occultior et aliquid satisfactio levitatis habere videatur. sed
viderimus. tu modo quam saepissime ad me aliquid.

Epistulam Lucceio quam misi, qua meas res ut scribat rogo, fac ut ab eo sumas (valde bella est) eumque ut adproperet adhorteris, et quod mihi se ita facturum rescripsit agas gratias, domum nostram quoad poteris invisas, Vestorio aliquid significes. valde enim est in me liberalis.

84 (IV. 10)

Scr. in Cumano ix Kal. Mai. an. 55 (§2)

⟨CICERO ATTICO SAL.⟩

1 Puteolis magnus est rumor [et] Ptolemaeum esse in regno. si
quid habes certius, velim scire. ego hic pascor bibliotheca
Fausti. fortasse tu putabas his rebus Puteolanis et Lucrinensibus.
ne ista quidem desunt, sed mehercule ⟨ut⟩ a ceteris oblectationi-
bus deseror et voluptat⟨ibus cum propter aetatem t⟩um propter
rem publicam, sic litteris sustentor et recreor maloque in illa
tua sedecula quam habes sub imagine Aristotelis sedere quam
in istorum sella curuli tecumque apud te ambulare quam cum
eo quocum video esse ambulandum. sed de illa ambulatione
2 fors viderit aut si quis est qui curet deus. nostram ambulationem
et Laconicum eaque quae circa sunt velim quo⟨ad⟩ poteris
invisas et urgeas Philotimum ut properet, ut possim tibi
aliquid in eo genere respondere.

3, 6 satisfactio ς: -s facito *ORMm*: -s facete (-ere) *Nbds*.
4, 3 quam *GNPs*: quidem *O*: numq- *R codd. Mal.* Δ: nunc q- Z^b.
7 liberalis *Ps*: -rtatis Ω.

Ep. 84] *novam ep. fecit Corradus; cum ep.* 85 (IV. 9) *coniungunt codd.* (*ambo om. O*).

1, 1 et *GR*Δ: *om. N*.
4 ista *NR*Δ: ist(a)e *EG*. ut *add. Bosius*.
5 et voluptatibus...tum *scripsi*: et voluptatum (-tibus *Hs*) *GN*ΔZ^bλ: voluptatum (-te *R*) *ER*.
10 quis *ENR*δ: qui *GM*λ.

Moreover the self-control I have shown in conduct might not be so apparent in writing, and such an apologia might seem
4 rather lacking in dignity. But we shall see. For your part, write me a line as often as you can.

Mind you get Lucceius to lend you the letter I sent him, asking him to write an account of my doings—a very pretty piece—and urge him to get on with it, and thank him for promising to do so in his reply. Please keep an eye on my house, as far as you can, and say something to Vestorius—he is behaving very handsomely to me.

84 (IV. 10)

Cumae, 22 April 55

CICERO TO ATTICUS

1 At Puteoli there is a strong rumour that Ptolemy is on his throne. If you have more reliable information, I should be glad to know. I am living here on Faustus' library—*you* perhaps think it's on these Puteolan and Lucrine commodities. Well, I have them too. But seriously, while all other amusements and pleasures have lost their charm because of my age and the state of our country, literature relieves and refreshes me. I would rather sit on that little seat you have underneath Aristotle's bust than in our Consuls' chairs of state, and I would rather take a walk with you at your home than with the personage in whose company it appears that walk I must. But as for *that* walk, chance and the gods, if any of them is interested,
2 must provide. As for *my* walk and my Laconian bath and its environs, I should be grateful if you would keep an eye on them as far as possible and urge Philotimus to make haste so that I can offer you some sort of return in this respect.

2, 2 circa *EGRδ* (curae *b*): cirea *M*: certa *N*. sunt *E*: sint *GNRΔ*. quoad (quod) *Graevius*: quo *GNRΔ*: cum *E*. poteris *Es*: -it *GNRΔ*.

Pompeius in Cumanum Parilibus venit. misit ad me statim qui salutem nuntiaret. ad eum postridie mane vadebam cum haec scripsi.

85 (IV. 9)

Scr. Neapoli iv Kal. Mai. an. 55 (§ 2)

CICERO ATTICO SAL.

1 Sane velim scire num censum impediant tribuni diebus vitiandis (est enim hic rumor) totaque de censura quid agant, quid cogitent. nos hic cum Pompeio fuimus. multum mecum de re publica, sane sibi displicens, ut loquebatur (sic est enim in hoc homine dicendum), Syriam spernens, Hispaniam iactans, hic quoque ut loquebatur — et opinor, usquequaque, de hoc cum dicemus, sit hoc quasi 'καὶ τόδε Φωκυλίδου'. tibi etiam gratias agebat quod signa componenda suscepisses; in nos vero suavissime hercule ⟨est⟩ effusus. venit etiam ad me in Cumanum a. d. VI. nihil minus velle mihi visus est quam Messallam consulatum petere. de quo ipso si quid scis velim scire.

2 Quod Lucceio scribis te nostram gloriam commendaturum et aedificium nostrum quod crebro invisis, gratum. Quintus frater ad me scripsit se, quoniam Ciceronem suavissimum tecum haberes, ad te Non. Mai. venturum. ego me de Cumano movi a. d. V Kal. Mai. eo die Neapoli apud Paetum. a. d. IIII Kal. Mai. iens in Pompeianum bene mane haec scripsi.

Ep. 85] 1, 1 impediant *R Ant.*: -endant *GNΔ*. diebus...petere (*v.* 11) *om. GH.*

3 multum *R*: -to *NΔ*: -ta *Ps.*

4 displicens *Rs*: -pic(i)ens *NΔ*.

5 iactans ς: l(a)etans Ω.

9 est ς: et Ω: *om.* Z^l.

10 a. d. VI *Constans*: at si Δ: si *NR*. velle ς: -em Ω.

2, 4 tecum haberes *Manutius*: secum haberet Ω.

5 Paetum *Corradus*: l(a)etum *GRΔ*: locum *N*.

Pompey arrived at his Cuman villa on Shepherds' Day, and at once sent a messenger to me with his compliments. That was yesterday, and I am just going over to him early to-day as I write this.

85 (IV. 9)

Naples, 27 April 55

CICERO TO ATTICUS

1 I should be very glad to know whether the Tribunes are holding up the census by voiding the days (there is a rumour here to that effect), and what they are doing and proposing to do about the Censorship generally. I have been with Pompey here. He discussed politics with me a good deal, not without much self-dissatisfaction from what he said (one has to put it that way in his case), scorning Syria, spurning Spain—again, from what he said. Indeed, every time we speak of Pompey I think it should be with this refrain, like 'This also says Phocylides'. He further spoke appreciatively of your undertaking to arrange his art collection. Towards myself I must say he was most agreeably effusive. He also paid a visit to my place at Cumae on the 25th. It seemed to me that the last thing he wants is for Messalla to stand for the Consulship. If you know anything on that subject by the way, I should like to know too.

2 Thank you for saying that you will recommend my glory to Lucceius and for keeping a watchful eye on my building. My brother writes to me that as you have our delightful young Quintus with you he will pay you a visit on the Nones of May. I left Cumae on 26 April and stayed the night at Naples with Paetus. I write this very early on 27 April, just off for Pompeii.

87 (IV. 13)

Scr. in Tusculano xvi vel xv Kal. Dec. an. 55 (§ 1)

CICERO ATTICO SAL.

1 Nos in Tusculanum venisse a. d. XVII Kal. Dec. video te scire.
ibi nobis Dionysius praesto fuit. Romae a. d. XIII Kal. volumus
esse. quod dico 'volumus', immo vero cogimur. Milonis
nuptiae. comitiorum non nulla opinio est. ego, etsi * * *,
⟨pro re⟩ nata afuisse me in altercationibus quas in senatu factas
audio fero non moleste. nam aut defendissem quod non place-
ret aut defuissem cui non oporteret. sed mehercule velim res
istas et praesentem statum rei publicae et quo animo consules
ferant hunc σκυλμὸν scribas ad me quantum pote. valde sum
2 ὀξύπεινος et, si quaeris, omnia mihi sunt suspecta. Crassum
quidem nostrum minore dignitate aiunt profectum paludatum
quam olim aequalem eius L. Paulum, item iterum consulem.
o hominem nequam!

De libris oratoriis factum est a me diligenter. diu multumque in manibus fuerunt. describas licet. illud etiam te rogo, τὴν παροῦσαν κατάστασιν τυπωδῶς, ne istuc hospes veniam.

Ep. 87] 1, 2 Dionysius nobis Δ. XIII *GRc*: XIIII *NO*: XVIII Δ.
3 quid *s*.
4 est...licet (2, 6) *paucis exceptis om. GH.* ego *Bosius*: ergo $O\delta Z^b$: rigo *NRM*.
4–5 etsi...pro re nata *scripsi* (*post* etsi *aliqua intercidisse viderat Fr. Schmidt*): et si irata *O*(?) *codd. Mal.* $bdsZ^b$: et sur- *NRMm*.
5 afuisse Z^l: abf- *s*: fuisse Ω (*sed post* me *R*).
9 scribas *NO*(?)*RC*: res- Δ. pote Ωλ: -es *bds*. valde *OFs*λ: val *NR*Δ.

87 (IV. 13)

Tusculum, 15 or 16 November 55

CICERO TO ATTICUS

1 I see you know that I arrived at Tusculum on 14 November.
Dionysius was there to meet me. I want to be in Rome on the
18th. I say 'want', but in fact I have to be there—Milo's
wedding. There is some expectation of elections. For my part,
although * * *, as things stand I am not sorry to have been
absent during the altercations which I hear have taken place in
the Senate, for I should either have had to support a measure I
did not like or to fail a man I ought to support. But I should
really be grateful if you would write to me as much as possible
about these matters, and the present political situation, and how
the Consuls are taking this *tracasserie*. I am quite greedily in-
quisitive, and, if it's of any interest to you, I am suspicious of
2 everything. They say that our friend Crassus left Rome in
uniform with rather less *éclat* than his coeval L. Paulus, also
Consul for the second time, in days gone by. What a rascal he
is!

I have not been idle over the work on oratory. It has been in my hands much and long. You can copy it. May I ask you again to give me the present situation *dans les grandes lignes*, so that I shall not come back to Rome like a foreigner?

2, 2 minore ς: -em Ω.
3 item iterum *Bücheler*: mit- it- *H*: item T. terum *et sim.* Ω.
5 diu multumque ς: diū ūttū que *M et sim. Rm*: diminutum qu(a)e *NObds*.
7 τυπωδῶς *vulg.*: ΤΥΠΟΔΩ *vel sim. GRM*. veniam *Oms*: -at *GNRMC*.

88 (IV. 14)

Scr. in Cumano vel Pompeiano med. m. Mai. an. 54

CICERO ATTICO SAL.

1 Vestorius noster me per litteras fecit certiorem te Roma a. d. VI Id. Mai. putare profectum esse tardius quam dixeras quod minus valuisses. si iam melius vales, vehementer gaudeo. velim domum ad te scribas ut mihi tui libri pateant non secus ac si ipse adesses, cum ceteri tum Varronis. est enim mihi utendum quibusdam rebus ex his libris ad eos quos in manibus habeo; quos, ut spero, tibi valde probabo.

2 Tu velim, si quid novi forte habes, maxime a Quinto fratre, deinde a C. Caesare, et si quid novi forte de comitiis, de re publica (soles enim tu haec festive odorari), scribas ad me; si nihil habebis, tamen scribas aliquid. numquam enim mihi tua epistula aut intempestiva aut loquax visa est. maxime autem rogo rebus tuis totoque itinere ⟨feliciter ex⟩ sententiaque confecto nos quam primum revisas. Dionysium iube salvere. cura ut valeas.

89 (IV. 16)

Scr. Romae c. Kal. Quint. an. 54

CICERO ATTICO SAL.

1 Occupationum mearum vel hoc signum erit quod epistula librari manu est. de epistularum frequentia te nihil accuso, sed pleraeque tantum modo mihi nuntiabant ubi esses quod erant abs te, vel etiam significabant recte esse. quo in genere maxime

---

Ep. 88] *In hac ep. sequentibusque plerumque deficiunt GH.*
1, 1 Roma ς: -am Ω.
2 putare possum (-sim *PF*) profectum *RF*. dixeras *Faërnus*: -at Ω.
2, 1 tu *VR*λ: ut *NMm*: *om. GP*. forte novi Δ.
6 feliciter ex *addidi* (ex ς). confecto ς: -ta Ω.
Ep. 89] **1,** 2 manu *s*: -us Σ*C*: in manus Δ.
3 plereque *M corr.*: -ique Ω (-unque *N*). erant *Malaespina*: erat Ω.

88 (IV. 14)

Cumae or Pompeii, mid-May 54

CICERO TO ATTICUS

1 Our friend Vestorius has written informing me that he thinks you left Rome on 10 May later than you had told me because you had not been well. If you are now better, I am very glad. I should be grateful if you would write to your house in Rome and tell them to give me free access to your books just as if you were at home yourself, Varro's among the rest. I need certain things in these books for the work I have in hand, which I hope you are going to like more than a little.

2 If you happen to have any news, from my brother Quintus in the first place and from C. Caesar in the second, or anything about the elections and politics (you are generally pretty good at getting wind of such things), I'd be grateful if you would write to me about it. If you haven't anything, write something or other all the same. Never have I found any letter of yours unseasonable or garrulous. But most of all let me beg you to come back to us as soon as possible, when you have finished your business and your whole trip prosperously and to your satisfaction. Give my regards to Dionysius. Take good care of yourself.

89 (IV. 16)

Rome, about 1 July 54

CICERO TO ATTICUS

1 The very fact that this letter is in a secretary's hand will show you how busy I am. I have no fault to find with you about the frequency of your letters, but most of them merely tell me your whereabouts in that they carry your address, or give the further information that you are well. I was particularly pleased by two of this sort, dispatched about the same time from

delectarunt duae fere eodem tempore abs te Buthroto datae; scire enim volebam te commode navigasse. sed haec epistularum frequentia non tam ubertate sua quam crebritate delectavit: illa fuit gravis et plena rerum quam mihi M. Paccius, hospes tuus, reddidit. ad eam rescribam igitur, et hoc quidem primum. Paccio et verbis et re ostendi quid tua commendatio ponderis haberet. itaque in intimis est meis, cum antea notus non fuisset.

2 Nunc pergam ad cetera. Varro, de quo ad me scribis, includetur in aliquem locum, si modo erit locus. sed nosti genus dialogorum meorum. ut in oratoriis, quos tu in caelum fers, non potuit mentio fieri cuiusquam ab iis qui disputant nisi eius qui illis notus aut auditus esset, ⟨ita⟩ hanc ego de re publica quam institui disputationem in Africani personam et Phili et Laeli ⟨et⟩ Manili contuli. adiunxi adulescentis Q. Tuberonem, P. Rutilium, duo Laeli generos, Scaevolam et Fannium. itaque cogitabam, quoniam in singulis libris utor prohoemiis ut Aristoteles in iis quos ἐξωτερικοὺς vocat, aliquid efficere ut non sine causa istum appellarem, id quod intellego tibi placere. utinam modo conata efficere possim! rem enim, quod te non fugit, magnam complexus sum et plurimi oti, quo ego maxime egeo.

3 Quod in iis libris quos laudas personam desideras Scaevolae, non eam temere dimovi, sed feci idem quod in πολιτείᾳ deus ille noster Plato. cum in Piraeum Socrates venisset ad Cephalum, locupletem et festivum senem, quoad primus ille sermo

---

1, 5 Buthroto λ: bruto Ω.
7 crebritate *Lambinus*: celebr- *GRMC* (-tem *R*): celeri- *NVP*δ.
10 Paccio *Lambinus* (?): actio *E*: actione *G*: pactione (-ni *R*) *VR*: ratione *N*: paccio (pact-) ratione Δ.

2, 1 includetur *GNVRC*: inducetur *Em*: *om. M.*
3 oratoriis *NP*: -ris *EVR*Δ.
4 mentio potuit Δ.
5 ita *add. Wesenberg.*
6–7 Phili et Laeli(i) *vulg.*: pilti et peti(i) *vel sim.* Ω.
7 et *P*: *om.* Ω.
8 duo *EVMC*λ: duos *NOR*δ.

your house at Buthrotum, for I wanted to know that you had had a good voyage. But this spate of letters has gratified me more by quantity than quality. Now the one you gave your guest M. Paccius really *was* a substantial epistle, full of matter. So I shall reply to that, and begin by saying that I have shown Paccius both in word and deed how heavily a recommendation from you counts. Accordingly he is in my inner circle, though I had never met him previously.

2 Now to the other points. Varro, of whom you write to me, shall have a place somewhere, provided there *is* a place. But you know the form of my dialogues. Just as in my work on Oratory, of which you speak so very handsomely, none of those taking part in the discussion could make mention of persons other than those they had known or heard, in the same way I have put this discussion on the State that I have embarked upon into the mouths of Africanus, Philus, Laelius and Manilius, with the addition of some young men, Q. Tubero, P. Rutilius, and Laelius' two sons-in-law, Scaevola and Fannius. So I am thinking of making a suitable occasion to address him in one of the prefaces which I am writing to each book, as Aristotle did in what he calls his 'exoteric' pieces. I understand that you would favour that. I only hope I can finish what I have undertaken. It's a big subject, as you realize, that I am grasping and an important one, requiring a great deal of time of which I am particularly short.

3 You say you regret Scaevola's disappearance from the work which you are good enough to praise. I did not drop him casually, but followed the example of our divine Plato in his 'Republic'. Socrates calls on Cephalus, a rich, genial old gentleman, in the Piraeus. During the opening talk the old

10 Aristoteles ς: adistolis *vel sim.* Ω. ἐξωτ- ς: ext- Ω.
13 quo *VR*δ: quod *GNM*.
3, 2 iddem (*sic*) quod in *E*: idem (id *bds*) in *GN*δ: id enim *M*(?): quod in *VR*.

habe[re]tur, adest in disputando senex; deinde, cum ipse quoque commodissime locutus esset, ad rem divinam dicit se velle discedere neque postea revertitur. credo Platonem vix putasse satis consonum fore si hominem id aetatis in tam longo sermone diutius retinuisset. multo ego magis hoc mihi cavendum putavi in Scaevola, qui et aetate et valetudine erat ea qua esse meministi et iis honoribus ut vix satis decorum videretur eum pluris dies esse in Crassi Tusculano. et erat primi libri sermo non alienus a Scaevolae studiis; reliqui libri τεχνολογίαν habent, ut scis. huic ioculatorem senem illum, ut noras, interesse sane nolui.

4 De re Piliae quod scribis, erit mihi curae. etenim est luculenta res Aureliani, ut scribis, indiciis, et in eo me etiam Tulliae meae venditabo. Vestorio non desum; gratum enim tibi id esse intellego et ut ille intellegat curo. sed scis qui; cum habeat duo facilis, nihil difficilius.

5 Nunc ad ea ⟨quae⟩ quaeris de C. Catone. lege Iunia et Licinia scis absolutum; Fufia ego tibi nuntio absolutum iri, neque patronis suis tam libentibus quam accusatoribus. is tamen et mecum et cum Milone in gratiam rediit. Drusus reus est factus a Lucretio. iudicibus reiciiendis ⟨dies⟩ a. d. v Non. Quint. de Procilio rumores non boni; sed iudicia nosti. Hirrus cum Domitio in gratia est. senatus consultum quod hi consules

---

3, 5 habetur *Boot*: -eretur Ω.
8 consonum Δ: commodum Σ.
9 magis *Ernesti*: satis *NVRMbd*: satius *Ems*.
10 qua *Rbds*: quam *ENVMm*.
14 ioculatorem Ω*C*λ: -ri(a)e δ.
4, 1 Piliae *Schütz*: fi- Ω.
2 indiciis *N*δ: iudici(i)s *GVRMZ*[l].
3 venditabo *MC*: -icabo Σδ.
5, 1 quae *Ps*: *om.* Ω.
2 Fufia *vulg.*: -i(a)e (fu sic *R*) Ω.
4 rediit *NObds*: redit *RMm*.

fellow is present at the discussion, but then, after speaking himself and very nicely too, he says he has to go and attend to a sacrifice, and does not reappear. I imagine that Plato thought it would not be quite *convenable* to keep a man of Cephalus' age too long in so protracted a conversation. I felt this consideration arose for me much more strongly in Scaevola's case, in view of his age and state of health, which were what you remember, and the eminence of his career, which made it seem hardly proper for him to be spending several days in Crassus' Tusculan villa. Moreover the conversation in Book I was sufficiently relevant to Scaevola's interests. The others, as you know, contain technicalities, in which I definitely did not want the old gentleman to participate—you remember how fond he was of his joke.

4 You write of Pilia's affair. I shall attend to it. From Aurelianus' reports it certainly seems to be of value, as you say. And I shall be making myself popular with my Tullia into the bargain. I am doing my best for Vestorius. I realize that this will please you and am taking care that *he* realizes it. But you know what he is—with two such easy-going fellows to deal with he is the most difficult of mortals.

5 Now to your inquiries about C. Cato. You know that he has been acquitted of charges under the lex Junia Licinia, and I tell you here and now that he will be acquitted of others under the lex Fufia, and his prosecutors will be happier about the result than his counsel. However, he has made his peace with both Milo and myself. Drusus has been charged by Lucretius; jury to be challenged on 3 July. There are sinister reports about Procilius, but you know the courts. Hirrus and Domitius are friends. The senatorial decree which the present Consuls have

---

5 dies *addidi* (dies est dictus *Madvig*).
6 Procilio *Manutius*: pr(a)elio Ω. nosti *NOδ*: -tri *RM*.
7 hi *Boot*: hic Ω.

de provinciis fecerunt, 'quicumque posthac...', non mihi vi||detur esse valiturum.

6 De Messalla quod quaeris, quid scribam nescio. numquam ego vidi tam paris candidatos. Messallae copias nosti. Scaurum Triarius reum fecit. si quaeris, nulla est magnopere commota συμπάθεια, sed tamen habet aedilitas eius memoriam non ingratam et est pondus apud rusticos in patris memoria. reliqui duo plebeii sic exaequantur [ut] Domitius ut valeat amicis, adiuvetur tamen non gratissimo munere, Memmius Caesaris commendetur militibus, Pompei gratia nitatur. quibus si non valuerit, putant fore aliquem qui comitia in adventum Caesaris detrudat, Catone praesertim absoluto.

7 Paccianae epistulae respondi; cognosce cetera. ex fratris litteris incredibilia quaedam de Caesaris in me amore cognovi, eaque sunt ipsius Caesaris uberrimis litteris confirmata. Britannici belli exitus exspectatur; constat enim aditus insulae esse muratos mirificis molibus. etiam illud iam cognitum est, neque argenti scrupulum esse ullum in illa insula neque ullam spem praedae nisi ex mancipiis; ex quibus nullos puto te litteris aut musicis eruditos exspectare.

8 Paulus in medio foro basilicam iam paene texerat isdem antiquis columnis. illam autem quam locavit facit magnificentissimam. quid quaeris? nihil gratius illo monumento, nihil gloriosius. itaque Caesaris amici, me dico et Oppium dirumparis licet, ⟨in⟩ monumentum illud quod tu tollere laudibus

---

5, 9 videtur *Mommsen*: ut (*sic etiam* Z^{b}λ) detur Ω, *ita tamen ut inter duas has voces archetypi foliis inter se scilicet mutatis verba quae sunt* quod iam intellegebamus (91. 3, 2) *usque ad* absolutum Gabinium (93. 1, 10) *inserant codd.; quae suo loco restituit idem* (*sed vide infra ad* 91. 3, 2). *Tornesianum itidem peccasse ex Lambini Bosiique adnotationibus procul est dubio.*

6, 2 nosti δ: notis *O*(*?*)*RM*: noras *N*.
6 plebeii ς: -bi Ω. ut *secl. Sternkopf.*
8 commendetur *vulg.*: -datur *s*: -dat Ω. gratia *s*: gallia Ω.

7, 5 muratos *NRbs*: mir- *OMdm*: munitos *E*.

8, 1 texerat ς: -rit ΩCZ^{l}λ: texit *bds*.
5 in *Bosius*: *om.* Ω$Z^{(b)}$.

passed about the provinces, 'Whosoever henceforth...', seems to me unlikely to work.

6 I don't know what to say to your question about Messalla. I have never seen candidates so evenly matched. Messalla's resources you know. Scaurus has been prosecuted by Triarius. I may add that no very noticeable sympathy has been aroused on his behalf, but still his Aedileship is not ungratefully remembered and his father's memory counts with the country voters. There remain the two plebeians, who are nicely balanced, Domitius having powerful friends and being helped by his show (not a particularly popular one however), while Memmius is recommended to Caesar's soldiers and relies on Pompey's influence. It is thought that if these advantages do not suffice, someone will be found to stave off the elections until Caesar comes, especially after Cato's acquittal.

7 The Paccius letter having been answered, let me tell you the rest of my news. A letter from my brother contains some quite extraordinary things about Caesar's warm feelings towards me, and is corroborated by a very copious letter from Caesar himself. The result of the war against Britain is eagerly awaited, for the approaches to the island are known to be 'warded with wondrous massy walls'. It is also now ascertained that there isn't a grain of silver on the island nor any prospect of booty apart from captives, and I fancy you won't expect any of *them* to be highly qualified in literature or music!

8 Paulus has now almost roofed his basilica in the middle of the Forum, using the original antique pillars. The other one, which he gave out on contract, he is constructing in magnificent style. It is indeed a most admired and glorious edifice. So Caesar's friends (I mean Oppius and myself, choke at that if you must) have thought nothing of spending sixty million sesterces on the work which you used to be so enthusiastic

solebas, ut forum laxaremus et usque ad atrium Libertatis explicaremus, contempsimus sescenties sestertium; cum privatis non poterat transigi minore pecunia. efficiemus rem gloriosissimam. nam in campo Martio saepta tributis comitiis marmorea sumus et tecta facturi eaque cingemus excelsa porticu ut mille passuum conficiatur. simul adiungetur huic operi villa etiam publica. dices 'quid mihi hoc monumentum proderit?' at quid id laboramus? ⟨habes⟩ res Romanas. non enim te puto de lustro, quod iam desperatum est, aut de iudiciis quae lege Clodia fiunt quaerere.

9 Nunc te obiurgari patere, si iure. scribis enim in ea epistula quam C. Decimius mihi reddidit Buthroto datam in Asiam tibi eundum esse te arbitrari. mihi mehercule nihil videbatur esse in quo tantulum interesset utrum per procuratores ageres an per te ipsum ut a nobis totiens et tam longe abesses. sed haec mallem integra re tecum egisse; profecto enim aliquid egissem. nunc reprimam susceptam obiurgationem. hoc quod dixi tantum utinam valeat ad celeritatem reditus tui!

Ego ad te propterea minus saepe scribo quod certum non habeo ubi sis aut ubi futurus sis; huic tamen nescio cui, quod videbatur is te visurus esse, putavi dandas esse litteras. tu quoniam iturum te in Asiam esse putas, ad quae tempora te exspectemus facias me certiorem velim, et de Eutychide quid egeris.

8, 6 forum *C*: eorum Ω. usque *N*: u (*cett. amissis*) *Frag. Angelomontanum* (*Ang.*): iisque *vel sim. ORΔ.*

7 sextercium (sest-) *NOMm*: cōs tercium *R*: tium (*prioribus amissis*) *Ang.*: HS *bds.*

11 simul *Nδ*: semel *ORM*(?) *Ang.*

12 dices *bds*: dicis *NORMm Ang.*

13 at *Wesenberg*: ad Ω *Ang.* quid id Δ: qui id *Ang.*: quid te *F*: quid *NOR.* labores *F.* habes *add. Lambinus.*

15 Clodia *Lange*: coctia *vel* coc(c)ia Ω. fiunt *Wesenberg:* fiant Ω *Ang.*

9, 4 interesset *bds*: -esse *NORMm Ang.*

5 ut a nobis *Baiter*: mutabis *NRΔ*: ut ab iis *F.*

6 egisse *NRδ*: -em *M.*

about, to widen the Forum and extend it as far as the Hall of Liberty. We couldn't settle with the previous owners for a smaller sum. We shall achieve something really glorious. As for the Campus Martius, we are going to build covered marble booths for the Assembly of Tribes and to surround them with a high colonnade, a mile of it in all. At the same time the Villa Publica will be attached to our building. You'll say, 'What good will such a structure be to me?' Now why should we worry ourselves about *that*? Well, there you have the news of Rome—I don't suppose you are interested in the census, which has now been given up as a bad job, or the trials under the lex Clodia.

9 Now submit to a scolding, if you deserve it. You say in the letter given from Buthrotum which C. Decimius brought me that you think you had better go to Asia. I really must say that I don't see where it makes such a difference whether you act through your agents or in person as to justify your being so long and so far away. But I wish I had tackled you about this before you made up your mind. I'm sure I should have made some impression. As it is I'll call a halt to my scolding. I only hope that what I *have* said may have the effect of hastening your return.

I write to you less often than I otherwise should because I am not certain of your present or future whereabouts. But I thought I ought to give a letter to this I don't know who because he seems likely to be seeing you. As you think you will go to Asia, I should be glad to know approximately when we may expect you back, and what you have done about Eutychides.

7–8 hoc...tantum $Z^b\lambda$: *om.* Ω.
8 tui ς: tuus Ω (*om. R*).
11 is te δ: isti Ω.

90 (IV. 15)

Scr. Romae vi Kal. Sext. an. 54 (§*8*)

CICERO ATTICO SAL.

1 De Eutychide gratum, qui vetere praenomine novo nomine T. erit Caecilius, ut est ex me et ex te iunctus Dionysius M. Pomponius. valde mehercule mihi gratum est Eutychidem tua erga me benevolentia cognosse suam illam in meo dolore συμπάθειαν neque tum mihi obscuram neque post ingratam fuisse.

2 Iter Asiaticum tuum puto tibi suscipiendum fuisse. numquam enim tu sine iustissima causa tam longe a tot tuis et hominibus et rebus carissimis et suavissimis abesse voluisses. sed humanitatem tuam amoremque in tuos reditus celeritas declarabit. sed vereor ne lepore suo detineat diutius praetor Clodius et homo pereruditus, ut aiunt, et nunc quidem deditus Graecis litteris Pituanius. sed si vis homo esse, recipe te ad nos ad quod tempus confirmasti. cum illis tamen cum salvi venerint Romae vivere licebit.

3 Avere te scribis accipere aliquid a me litterarum. dedi, ac multis quidem de rebus ἡμερολεγδὸν perscripta omnia; sed, ut conicio, quoniam mihi non videris in Epiro diu fuisse, redditas tibi non arbitror. genus autem mearum ad te quidem litterarum eius modi fere est ut non libeat cuiquam dare nisi de quo exploratum sit tibi eum redditurum.

4 Nunc Romanas res accipe. a. d. III⟨I⟩ Non. Quint. Sufenas et Cato absoluti, Procilius condemnatus. ex quo intellectum est

---

Ep. 90] 1, 3 gratum est (-mst) *Bücheler*: -m si *NΔ*: -m se *O(?)R*.
3–4 tua...benevolentia *Baiter*: -am...-am Ω.
4 cognosse suam *Baiter*: -osces suam *N*: -ossi iam *O* (-sse?) *Mm*: -oscam *R*: -osses iam *bds*.
2, 4 declarabit ς: -avit Ω.
7 Pituanius *Manutius*: -ios Ω.
8 tamen *GRM*: tum *NPms*.
3, 1 avere *ms*: habere Ω.
5 fere *R*: forte *GNΔ*.

90 (IV. 15)

Rome, 27 July 54

CICERO TO ATTICUS

1 Thank you for the news about Eutychides. So he will be T. Caecilius in future, the old forename with the new name, just as Dionysius is M. Pomponius, made up of me and you. I am really very much pleased that Eutychides has found out, through your kindness to me, that I noticed his sympathy in my dark days and have not forgotten it since.

2 I think your trip to Asia must have been necessary, for you would never without the best of reasons have wished to be so far away from all the people and things that are yours and that you most love and delight in. But your goodness of heart and affection for those near to you will be plain to see in the speed of your return. I'm afraid though that Praetor Clodius' wit and charm may detain you a long while, and Pituanius too, such a scholar, they tell me, and so devoted nowadays to Greek literature! But be a good fellow and come back to us by the time you promised. After all you will be able to see plenty of them in Rome when they are safely home again.

3 You say you are longing to get a line from me. I *did* write, with a full day-by-day account of all sorts of things. But as I suspect, since it looks as though you did not stay long in Epirus, I suppose the letter was never delivered. My letters to *you* being of the kind they generally are, I don't like giving them to anybody unless I can be sure that he will deliver them to you.

4 Now let me give you the news of the town. On 4 July Sufenas and Cato were acquitted, Procilius found guilty. We conclude that our triple-piled Daniels don't care a rap about

---

4, 1 romanas δ: -nus *ORM*: romam *N*. accipe *N*δ: accepit *ORM*. IIII *Boot* (*coll. Ascon. 18.17*): III Ω.
2 condemnatus ς: com(m)endatus *M* (*in marg.*) *ms*: -atur *ORMbd*: -andus *N*.

τρισαρειοπαγίτας ambitum, comitia, interregnum, maiestatem, totam denique rem publicam flocci non facere, patrem familias domi suae occidi nolle, neque tamen id ipsum abunde; nam absolverunt XXII, condemnarunt XXVIII. Publius sane diserto epilogo ⟨la⟩crim[in]ans mentis iudicum moverat. Hortalus in ea causa fuit cuius modi solet. nos verbum nullum; verita est enim pusilla, quae nunc laborat, ne animum Publi offenderet.

5 His rebus actis Reatini me ad sua Τέμπη duxerunt ut agerem causam contra Interamnatis apud consules et decem legatos, qua lacus Velinus a M'. Curio emissus interciso monte in Nar⟨em⟩ defluit; ex quo est illa siccata et umida tamen modice Rosea. vixi cum Axio; quin etiam me ad Septem Aquas duxit.

6 Redii Romam Fontei causa a. d. VII Id. Quint. veni spectatum, primum magno et aequabili plausu — sed hoc ne curaris, ego ineptus qui scripserim. deinde Antiphonti operam. is erat ante manu missus quam productus. ne diutius pendeas, palmam tulit. sed nihil tam pusillum, nihil tam sine voce, nihil tam... verum haec tu tecum habeto. in Andromacha tamen maior fuit quam Astyanax, in ceteris parem habuit neminem. quaeris nunc de Arbuscula. valde placuit. ludi magnifici et grati; venatio in aliud tempus dilata.

7 Sequere nunc me in campum. ardet ambitus. 'σῆμα δέ τοι ἐρέω': faenus ex triente Id. Quint. factum erat bessibus. dices

---

4, 3 τρισ *Bosius*: tres *N*Δ: est *R*.
5 familias *bds*: -ia *ORMm*: -ie *N*. occidi *Schütz*: -dere Ω.
7 diserto ς: deserte Ω. lacrimans *scripsi*: cruminar *RM*: criminans *N*: criminari δ.
8 Hortalus *Manutius*: hotal(l)us *vel sim.* Ω: hortensius *s*.
9 verita *Manutius*: -tas ΩZ^b. publii *Nc*: -io *R*ΔZ^b.
5, 1 Reatini ς: erat in (i)me Ω. duxerunt ς: dix- Ω.
2 Interamnates ς: inamnatis *vel sim.* Ω. consules *scripsi*: -Ω.
3 qua *scripsi*: quod Ω. M'. *Manutius*: M. Ω.
4 Narem *Lambinus*: nar Ω.
5 quin *etiam C*.
6, 1 spectatum *Graevius*: -aculum Ω.

bribery, or elections, or interregnum, or lèse-majesté or indeed the entire body politic, but when it comes to slaughtering a paterfamilias in his own house, they disapprove—not, however, overwhelmingly even then; 22 voted 'Not Guilty' to 28 'Guilty'. The jury was moved by a really eloquent peroration from Publius, delivered with tears. Hortalus was in his usual form. I said not a word—my little girl, who is not very well at present, was afraid it might annoy Publius.

5 All this transacted, the good folk of Reate brought me to their Vale, wishing me to plead their case against the Interamnates before the Consuls and ten assessors. There Lake Velinus flows into the Nar through a cutting in the mountains made by M'. Curius, as a result of which the Plain of Rosea was drained, though remaining moderately moist. I stayed with Axius. Indeed he took me to see the Seven Waters.

6 I returned to Rome for Fonteius' benefit on 9 July, and went to the theatre. To begin with, the applause was loud and steady as I entered—but never mind that, I am a fool to mention it. To proceed, I saw Antipho, who had been given his freedom before they put him on the stage. Not to keep you too long in suspense, he won the prize; but never have I seen such a weedy little object, not a scrap of voice, not a—but don't say I said so! As Andromache at any rate he stood head and shoulders above Astyanax. In the other roles he didn't have his equal. Now you'll want to know about Arbuscula. First-rate. The games were fine and much appreciated. The hunt was put off to another time.

7 Now follow me to the Campus. Bribery is running riot. 'A token I shall tell': interest went up on the Ides of July from

---

3 is erat *Victorius*: miserat Ω.
5 *post* tam *aposiopesin indicavit Bosius.*
7 Astyanax, in *Victorius*: astra (-tia, -tya) nam in Ω: Astyanax. nam in *c*.
9 dilata ς: dillata *N*: delata *GNRΔ*.

'istuc quidem non moleste fero.' o virum! o civem! Memmium Caesaris omnes opes confirmant. cum eo Domitium consules iunxerunt, qua pactione epistulae committere non audeo. Pompeius fremit, queritur, Scauro studet; sed utrum fronte an mente dubitatur. ἐξοχὴ in nullo est; pecunia omnium dignitatem exaequat. Messalla languet, non quo aut animus desit aut amici, sed coitio consulum ⟨et⟩ Pompeius obsunt. ea comitia puto fore ut ducantur. tribunicii candidati iurarunt se arbitrio Catonis petituros. apud eum HS quingena deposuerunt ut, qui a Catone damnatus esset, id perderet et competitoribus tribueretur.

8 Haec ego pridie scribebam quam comitia fore putabantur. sed ad te v Kal. Sext., si facta erunt et tabellarius non erit profectus, tota comitia perscribam. quae si, ut putantur, gratuita fuerint, plus unus Cato potuerit quam ⟨omnes leges⟩ omnesque iudices.

9 Messius defendebatur a nobis de legatione revocatus; nam cum Caesari legarat Appius. Servilius edixit ut adesset. tribus habet Pomptinam, Velinam, Maeciam. pugnatur acriter; agitur tamen satis. deinde me expedio ad Drusum, inde ad Scaurum. parantur orationibus indices gloriosi. fortasse accedent etiam consules designati. in quibus si Scaurus non fuerit, in hoc iudicio valde laborabit.

7, 3 istuc *vulg.*: -um Ω.
5 consules *vulg.*: cons. *GR*: -sulem *NOP*Δ. epist(u)l(a)e *GOR*: -lam *N*Δ.
7 nullo *codd. Mal. bds*λ: -la Ω.
9 consulum *vulg.*: -lis Ω. et *add.* ς. obsunt *s*: abs- (aps-) *ORMm*: absens est *Nbd*. ea *Manutius*: ex *R*Δ: *om. N.*
12 esset ς: est Ω. et ς: ea *N*: a *R* (*sed* perdere): et a Δ.
8, 1 fore *ORs*: in f- *NMm*: in foro *bd*.
2 sed Δ: si *OR*: *om. N.* te si v *N.* si *om.* Σ.
4 omnes leges omnesque *Wesenberg*: omnes qui Ω.

$\frac{1}{3}$ to $\frac{2}{3}$%. 'That I can bear', you'll say—you model of morality and public spirit! Caesar is backing Memmius for all he's worth. The Consuls have paired him with Domitius in an arrangement which I dare not commit to writing. Pompey is fuming and growling. Ostensibly Scaurus is his man, but does he mean it? We have our doubts. None of the four has any special prominence. Money levels standing all round. Messalla is lagging, not for lack of spirit or friends, but the Consuls' pact and Pompey are against him. I think these elections are likely to drag on. The tribunician candidates have taken an oath to conduct their campaign with Cato as umpire. Each of them has deposited HS 500,000 with him on the understanding that anyone found guilty of impropriety by Cato shall forfeit his deposit, which will be distributed among his rivals.

8 I write this the day before the elections are expected to take place. If they do, and if the courier has not left, I shall write you the whole story of the elections on 28 July. If they go through without bribery, as it is believed they will, Cato will have done more single-handed than all the laws and all the jurymen.

9 I am defending Messius, who has been recalled from his Commissionership—Appius had given him one on Caesar's staff. Servilius issued an edict requiring him to attend. He has for tribes the Pomptine, the Veline, and the Maecian. We are making a game fight of it, but we have all we can do. Then I get ready for Drusus, from him to Scaurus. Fine titles for my speeches I'm collecting! Perhaps the Consuls-Designate will come next. If Scaurus is not one of them, he'll have a difficult passage in this trial.

9, 1 de Σ: sed de Δ.
2 legarat ϛ: legerat *ORΔC*: legaret *N*: -averat *s*.
3 Velinam, Maeciam *vulg.*: villam m(a)ecia *vel sim.* Ω.
5 accedent ϛ: accid- *N*Δ: occid- *OR*.

10 Ex Quinti fratris litteris suspicor iam eum esse in Britannia. suspenso animo exspecto quid agat. illud quidem sumus adepti, quod multis et magnis indiciis possumus iudicare, nos Caesari et carissimos et iucundissimos esse. Dionysium velim salvere iubeas et eum roges ut te hortetur quam primum venias, ut possit Ciceronem meum atque etiam me ipsum erudire.

91 (IV. 17)

Scr. Romae Kal. Oct. an. 54 (§4)

CICERO ATTICO SAL.

1 Puto te existimare me non oblitum consuetudinis et instituti mei rarius ad te scribere quam solebam; sed cum loca et itinera tua nihil habere certi video, neque in Epirum neque Athenas neque in Asiam neque cuiquam nisi ad te ipsum proficiscenti dedi litteras. neque enim ⟨eae⟩ sunt epistulae nostrae quae si perlatae non sint nihil ea res nos offensura sit; quae tantum habent mysteriorum ut eas ne librariis quidem fere committamus ne quid quo excidat.

2 Consules flagrant infamia quod C. Memmius candidatus pactionem in senatu recitavit quam ipse ⟨et⟩ suus competitor Domitius cum consulibus fecisset, uti ambo HS |XXXX| consulibus darent, si essent ipsi consules facti, nisi tris augures dedissent qui se adfuisse dicerent cum lex curiata ferretur quae lata non esset, et duo consularis qui se dicerent in ornandis provinciis consularibus scribendo adfuisse cum omnino ne senatus quidem

---

10, 5 ut te hortetur (-entur *G*) *GNΔ*: *om. OR.* veniat *OP.*

Ep. 91] 1, 2 mei *ERC*: tui me *GNVΔ* (mei me *b*, tui *s*): tui (*tunc praecisum*) *Ang.*

5 eae *add. Lambinus.*

8 ne quid quo *Schmidt*: lepidum quo ΩZ^b.

2, 2 et *add. Manutius.* competitor ς: rep- Ω *Ang.*

10 From a letter of my brother's I guess that he is already in Britain. I am waiting in suspense to hear how he is getting on. One thing has been achieved anyhow, as I can judge by many substantial tokens—I stand very high in Caesar's regard and favour. Remember me to Dionysius, and ask him to persuade you to come back as soon as possible, so that he can instruct my Marcus and myself too.

91 (IV. 17)

Rome, 1 October 54

CICERO TO ATTICUS

1 If I write to you less often than I used, I feel sure you won't think it is because I have forgotten my old-established habit. But as your addresses and routes seem to be quite uncertain, I have given no letters to travellers to Epirus or Athens or Asia or to anyone at all unless he is on his way to you personally. Mine are not the sort of letters which can miss their destination and no harm done. They contain so many secrets that I don't usually trust them even to my clerks for fear something might leak out.

2 The Consuls are in the thick of a tremendous scandal. The candidate C. Memmius read out in the Senate the text of a compact which he and his fellow-candidate Domitius had made with the Consuls providing that, if they were themselves elected, they would furnish three Augurs who would state that they had been present at the passing of a *lex curiata* which had never been passed, and three Consulars who would state that they had witnessed a senatorial decree making provision for the consular provinces, though the Senate had not so much as met; failing this, they were to forfeit the sum of HS 4,000,000

3 uti ambo ς: utinambo *Ang.*: -nam ambo Δ: -nam bonus (HS *om.*) *R*: -nam bona *V*: uti. nam *N*. |xxxx| *scripsi*: quadragena Ω*C*.

fuisset. haec pactio non verbis sed nominibus et perscriptionibus multorum ⟨per⟩ tabulas cum esse facta diceretur prolata a Memmio est nominibus inductis auctore Pompeio. hic Appius erat idem; nihil sane iacturae. corruerat alter et plane, inquam,

3 iacebat. Memmius autem dirempta coitione invito Calvino plane refrixerat et eo magis nunc || quod iam intellegebamus enuntiationem illam Memmi valde Caesari displicere. Messalla noster et eius Domitius competitor liberalis in populo valde fuit. nihil gratius. certi erant consules. at senatus decernit ut tacitum iudicium ante comitia fieret ab iis consiliis †quae erant omnibus sortita† in singulos candidatus. magnus timor candidatorum. sed quidam iudices, in his Opimius, Veiento, Rantius, tribunos pl. appellarunt ne iniussu populi iudicarent. res ce⟨ci⟩dit. comitia dilata ex senatus consulto dum lex de tacito iudicio ferretur. venit legi dies. Terentius intercessit. consules, qui illud levi bracchio egissent, rem ad senatum detulerunt. hic Abdera, non tacente me. dices 'tamen tu non quiescis?' ignosce, vix possum. verum tamen quid tam ridiculum? senatus decreverat ne prius comitia haberentur quam lex lata esset; ⟨si⟩ quis intercessisset, res integra referretur. coepta ferri leviter, intercessum non invitis, res ad senatum. 'de ea re ita censuerunt, comitia primo quoque tempore haberi. c.'

2, 8 adfuisset Δ.
9 per tabulas *scripsi*: tabulas Ω: -lis *bs*.
3, 1 dirempta ς: -epta Ω. invito etiam calvino *VR*.
2 nunc *NRΔ*Z^b: in *V*. *sequentes litteras quae sunt* cociace *a Mommsenio hic positas ad* 93. 1, 10 *pertinere vidit Sternkopf*.
3 displicere *bs*: -rem *N*: despicerem *ORMdm*.
5 erant ς: erat Ω. at ς: ac Ω.
9 Veiento, Rantius *Bücheler*: Veientor- Z^b: veietor Antius *C*: veietora- *vel sim.* Ω.
10 res cecidit *Madvig*: rescidit (res c- *M*) *NOPΔ*Z^l: res cedit *Ant.*: recidie *R*.
11 dum ς: cum Ω.
13 retulerunt *s*.
16 si quis *Sjögren, auctore Malaespina*: quis Σ: qui si Δ.
17 ferri ς: fieri Ω.
19 c. *scripsi*: censere Ω.

to the Consuls. The compact, which, as was stated, was not an oral one but had been entered in the account books of many persons with names and full details, was produced by Memmius on Pompey's advice with the names cancelled. This exposure leaves Appius where he was—nothing lost to be sure. But for his colleague it is a collapse, he is utterly discredited.

3 Memmius, having broken up the coalition against Calvinus' wishes, looks a very tired horse, especially at this present, because we now gather that Caesar is very much annoyed at his making this disclosure. Our friend Messalla and his fellow-candidate Domitius have treated the public very handsomely. Much popularity in consequence. They were certain of election. Then came a senatorial decree providing that before the elections a 'silent' process should be instituted against each candidate individually, the jury to be made up by lot from those already empanelled (?). Great alarm among the candidates. Certain jurors however, including Opimius, Veiento and Rantius, appealed to the Tribunes against being required to serve without the authority of the Assembly. Collapse of project. The elections were then postponed by senatorial decree to allow time for a law concerning the 'silent' process to be submitted to the Assembly. Voting day arrived, Terentius cast his veto. The Consuls, who had handled the business very gingerly, referred the position to the Senate. All Fools Festival there, not without a protest from me. '*Still* not keeping quiet?' you'll say. Sorry, I scarcely know how to. But really, is it not too fantastic? The Senate had forbidden elections until the presentation of the law, with the proviso that in the event of a veto the matter should be referred to them afresh. The machinery is set in motion, half-heartedly enough, the veto received without reluctance, the matter brought back to the House. 'Concerning this matter this House declared that the elections be held as soon as possible. Declared.'

4 Scaurus, qui erat paucis diebus illis absolutus, cum ego partem eius ornatissime defendissem, obnuntiationibus per Scaevolam interpositis singulis diebus usque ad pr. Kal. Oct., quo ego haec die scripsi, sublatis, populo tributim domi suae satis fecerat. sed tamen, etsi uberior liberalitas huius, gratior esse videbatur eorum qui occuparant. cuperem vultum videre tuum cum haec legeres; nam profecto spem habes nullam haec negotia multarum nundinarum fore. sed senatus hodie erat futurus, id est Kal. Oct.; iam enim luciscit. ibi loque[re]tur praeter Ateium et Favonium libere nemo; nam Cato aegrotat. de me nihil timueris; sed tamen promitto nihil.

5 Quid quaeris aliud? iudicia, credo. Drusus, Scaurus non fecisse videntur. tres candidati rei fore putabantur, Domitius a Memmio, Messalla a Q. Pompeio Rufo, Scaurus a Triario aut a L. Caesare. 'quid poteris' inquies 'pro iis dicere?' ne vivam ⟨si⟩ scio; in illis quidem tribus libris quos tu dilaudas nihil reperio.

92 (iv. 18)

Scr. Romae inter ix Kal. et iv Non. Nov. an. 54 (§§ 4, 5)

⟨CICERO ATTICO SAL.⟩

1 *** nunc, ut opinionem habeas †rerum†, ferendum est. quaeris ego me ut gesserim. constanter et libere. 'ille' inquies 'ut ferebat?' humaniter, meaeque dignitatis, quoad mihi satis factum esset, habendam sibi rationem putabat. quo modo ergo

---

4, 1 partem *Manutius*: patrem *ORΔ*: pater *N*.
4 die scripsi *vulg.*: des- Ω.
5 gratior *Victorius*: er(r)atior *vel sim.* Ω.
6 vultum δ: ut cum (tn̄ *N*) Ω.
7 spem *Orelli*: rem Ωλ.
8 erat *N*: fuerat *ORΔ*.
9 luciscit. ibi loquetur *Victorius*: -ci tibi loqueretur Ω.
10 Ateium *scripsi*: antium Ω. libere Σ*C*: liber Δ.
5, 2 fore rei Δ.
5 si *add.* ς. quos tu dilaudas *Victorius*: quo (quos *R*) studio l- Ω*C*.

4 A few days ago Scaurus was acquitted, part of him being defended by me in my best style. Every day up to 29 September, the day I write this letter, having been voided by Scaevola with reports of bad auspices, he has given the public its due at his own house, tribe by tribe. But though his largesse has been more plentiful, it looks as if more popularity will go to those who got in first. I should love to see your face as you read this —for you really can't expect these goings-on to last many weeks. However, the Senate is to meet to-day, i.e. the Kalends of October, for it's now getting light. No one will speak his mind except Ateius and Favonius—Cato's ill. As for me, don't be nervous. Still, I make no promises.

5 What else would you care to hear about, I wonder? The trials, perhaps. Well, Drusus and Scaurus are found not guilty. Three of the candidates are expected to have to face a prosecution, Domitius by Memmius, Messalla by Q. Pompeius Rufus, Scaurus by Triarius or L. Caesar. What am I going to say for them, you may ask? I'll be hanged if I know. Certainly I find nothing in those three volumes which you praise so generously.

92 (IV. 18)

Rome, between 24 October and 2 November 54

CICERO TO ATTICUS

1 ...As it is, to give you my (?) view of things, we must grin and bear it. You ask how I comported myself. Firmly and frankly. 'Well, how did he take it?' Pleasantly enough—he seemed conscious of his obligation to consider my dignity until amends had been made to me. You wonder how Gabinius was

---

Ep. 92] 1, 1 *verba* nunc ut...sensibus (2, 18) ἀποσπασμάτιον *novae esse epistulae suspicatus est Malaespina, sequentia eodem pertinere ostendit Mommsen. codices cum superiore iungunt.* nunc...rerum *ita etiam* $Z^{(b)}$.
2 gesserim *Pbs*: ce- Ω.

absolutus? omnino πρῷρα πρύμνα accusatorum incredibilis infantia, id est L. Lentuli L. f., quem fremunt omnes praevaricatum, deinde Pompei mira contentio, iudicum sordes. ac tamen XXXII condemnarunt, XXXVIII absolverunt. iudicia reliqua impendent. nondum est plane expeditus.

2 Dices 'tu ergo haec quo modo fers?' belle mehercule et in eo me valde amo. amisimus, mi Pomponi, omnem non modo sucum ac sanguinem sed etiam colorem et speciem pristinam civitatis. nulla est res publica quae delectet, in qua acquiescam. 'idne igitur' inquies 'facile fers?' id ipsum. recordor enim quam bella paulisper nobis gubernantibus civitas fuerit, quae mihi gratia relata sit. nullus dolor me angit unum omnia posse; dirumpuntur ii qui me aliquid posse doluerunt. multa mihi dant solacia, nec tamen ego de meo statu demigro, quaeque vita maxime est ad naturam ad eam me refero, ad litteras et studia nostra. dicendi laborem delectatione oratoria consolor. domus me et rura nostra delectant. non recordor unde ceciderim sed unde surrexerim. fratrem mecum et te si habebo, per me isti pedibus trahantur; vobis ἐμφιλοσοφῆσαι possum. locus ille animi nostri stomachus ubi habitabat olim concalluit. privata modo et domestica nos delectent, miram securitatem videbis; cuius plurimae mehercule partes sunt in tuo reditu. nemo enim in terris est mihi tam consentientibus sensibus.

1, 5 πρῷρα πρύμνα *Manutius*: ΠΟΡΠΑΠΥΜΝΑ *vel sim. RMmZ^bc.*
6 f. *ms*: T. *vel* t Ω.
7–8 ac tamen *Wesenberg*: att- Ω.

2, 1 fers. belle mehercule *Ec* (*sed* ? *c*): fers (fert *N*, ferre Δ) vellem h- (me h- *N*) *NOR*Δ.
2 omnem non modo Σ*cZ^b*λ: omnem *M*: non modo o- δ.
4 a(c)quiescam *EV*δ: -at *GNM*: -as *R*.
5 fers *EVP*: fer *G*: fert *N*: fieri *R*: ferri Δ.
10 ad litteras *Lambinus*: et l- *EGNOR*Δ: id est l- *V*.
14 isti *Gulielmius*: ista Ω. trahantur *Malaespina*: -huntur Ω.
15 habitabat Ω: -tat *Nonius 1, p. 128* (*Lindsay*). concalluit ς: -aluit Ω, *Nonius*.
18 in terris *EV*: ΔΕΡΡΙϹ (δέρρις) in terris *vel sim. GRMm* (*cf.* 93 (IV. 19). 1): in t- (*praeced. spat.*) *Hbds*: in ceteris (*praeced. lin. distort.*) *N*.

acquitted? To be sure, the long and the short of it was the unbelievable oratorical incompetence of the prosecution, i.e. of L. Lentulus junior, who is loudly and universally accused of collusion. Add Pompey's extraordinary efforts and a corrupt jury. Even so 32 voted 'Guilty' to 38 'Not Guilty'. The other trials are still hanging over him. He's not yet completely out of the wood.

2 You'll wonder how I take all this. Pretty coolly, I assure you, and I plume myself highly on doing so. My dear friend, not only have we lost the vital essence of the free state—even the outward complexion and aspect it used to wear has gone. There is no Republic any longer to give me joy and solace. Can I take that calmly? Why yes, I can. You see, I have the memory of the proud show she made for the short time that I was at the helm, and the thanks I got in return. My withers are unwrung by the spectacle of one man all-powerful, which chokes the persons who found it distasteful that *I* should have any power at all. I have many consolations. All the same, I do not move away from my position, but turn back to the life that is most congenial, to my books and studies. The labour of pleading is compensated by the pleasure that oratory gives me. My house in town and my places in the country are a source of delight. I do not remember the height from which I fell but the depth from which I have risen. If I can have my brother's company and yours, then so far as I am concerned these people can go to the devil. I can philosophize and you can listen. That place in my mental anatomy which used to contain my spleen grew a tough skin long ago. Providing only that my private and domestic circumstances give me pleasure, you will find my equanimity quite remarkable. It largely depends, believe me, on your return. There is no one in the world with whom I hit it off quite so happily.

3 Sed accipe alia. res fluit ad interregnum et est non nullus odor dictaturae, sermo quidem multus; qui etiam Gabinium apud timidos iudices adiuvit. candidati consulares omnes rei ambitus. accedit etiam Gabinius, quem P. Sulla, non dubitans quin foris esset, postularat, contra dicente et nihil obtinente Torquato. sed omnes absolventur, nec posthac quisquam damnabitur nisi qui hominem occiderit. hoc tamen agitur severius, itaque indicia calent. M. Fulvius Nobilior condemnatus est; multi alii urbani ne respondent quidem.

4 Quid aliud novi? etiam. absoluto Gabinio stomachantes alii iudices hora post Antiochum Gabinium nescio quem ⟨e⟩ Sopolidis pictoribus libertum, accensum Gabini, lege Papia condemnarunt. itaque dixit statim †res p. lege maiestatis ουσοιμρισαμαφιηι†. Pomptinus vult a. d. IIII Non. Nov. triumphare. huic obviam Cato et Servilius praetor⟨es⟩ ad portam et Q. Mucius tribunus. negant enim latum de imperio, et est latum hercule insulse. sed erit cum Pomptino Appius consul. Cato tamen adfirmat se vivo illum non triumphaturum. id ego puto ut multa eiusdem ad nihil⟨um⟩ recasurum. Appius sine lege suo sumptu in Ciliciam cogitat.

5 A Quinto fratre et a Caesare accepi a. d. VIIII Kal. Nov. litteras datas a litoribus Britanniae proximae a. d. VI Kal. Oct. confecta Britannia, obsidibus acceptis, nulla praeda, imperata tamen pecunia exercitum e Britannia reportabant. Q. Pilius erat iam ad Caesarem profectus. tu, si aut amor in te est

---

3, 1 fluit *Manutius*: fuit Ω.
8 indicia ΩZ^l: iud- *Pms.* nobilior *N*δ: -ius *VRM.*
4, 1 etiam *Acidalius*: et tamen Ω.
3 e *Bosius*: in *R*: *om. NV*Δ. Sopolidis *vel* rop- Z^b: rop- Z^l: solidis Ω.
6 praetores *Corradus*: praetor Ω. ad portam *EGVR*: a porta *NMm.*
7 negant *EVR*: -at *GN*Δ.
9 triumphare *Mm.*
10 nihilum *Siesbye*: nihil Ω.
5, 2 datas...Oct. *huc transp. Klotz*: pecunia *sequuntur in codd.* proximae *scripsi*: -mo Ω.
4 e *R Ant.*: ē *N*: ex *VPF*: a *c*: *om. E*Δ.

3 Well, let me tell you of other matters. The situation is drifting towards an Interregnum and there is some whiff of a Dictatorship in the air, much talk anyhow—incidentally that helped Gabinius with the less courageous members of the jury. All the candidates for the Consulship are being prosecuted for corruption. Gabinius now joins them. P. Sulla, in the confidence that all was up with him, had brought the charge, Torquatus unsuccessfully contesting. But they'll all be let off, and nobody henceforward will ever be found guilty unless he's killed someone. This is still treated with some severity, so the informers are busy. M. Fulvius Nobilior has been convicted. Many others do not have the politeness to answer to their names.

4 What other news? Oh yes, within an hour of Gabinius' acquittal, a freedman in his service, one Gabinius Antiochus, a pupil of Sopolis the painter, was found guilty by another jury under the lex Papia, in sheer irritation at the result. Said the prisoner right away, 'Poor law of lèse-majesté! "Thy portion is not with the Paphian"' (?). Pomptinus wants to celebrate his triumph on 2 November. Praetors Cato and Servilius are stopping him at the city gate with Tribune Q. Mucius. They say that no enabling law has been brought in, and it's true that the manner of its bringing in was less than tactful. But Consul Appius will be on Pomptinus' side. Cato, however, asseverates that he will triumph over his (Cato's) dead body—an utterance which I fancy will end in air like a good many from the same quarter. Appius plans to go to Cilicia at his own expense and do without a law.

5 On 24 October I received letters from my brother Quintus and from Caesar dispatched 'Shores of Nearer Britain, 25 September'. They had settled Britain, taken hostages but no booty (tribute, however, imposed), and were bringing back the army from the island. Q. Pilius has already left to join Caesar. As for yourself, if you have any affection for your

nostri ac tuorum aut ulla veritas, aut si etiam sapis ac frui tuis commodis cogitas, adventare et prope adesse iam debes. non mehercule aequo animo te careo; te autem quid mirum, qui Dionysium tanto opere desiderem? quem quidem abs te, cum dies venerit, et ego et Cicero meus flagitabit. abs te proximas litteras habebam Epheso a. d. v Id. Sext. datas.

93 (IV. 19)

Scr. Romae ex. m. Nov. an. 54

CICERO ATTICO SAL.

1 O exspectatas mihi tuas litteras! o gratum adventum! o constantiam promissi et fidem miram! o navigationem amandam! quam mehercule ego valde timebam recordans superioris tuae transmissionis δέρρεις. sed, nisi fallor, citius te quam scribis videbo. credo enim te putasse tuas mulieres in Apulia esse; quod cum secus erit quid te Apulia moretur? an Vestorio dandi sunt dies et ille Latinus Ἀττικισμὸς ex intervallo regustandus? quin tu huc advolas et invisis illius nostrae rei publicae germanae puta⟨mina⟩? vide nummos ante comitia tributim uno loco divisos palam, vide absolutum Gabinium, || olface
2 dictaturam, fruere iustitio et omnium rerum licentia, perspice aequitatem animi mei et ludum et contemptionem Selicianae unciae et mehercule cum Caesare suavissimam coniunctionem;

5, 6 nostri *NF*: -ter *VP*: non *GR*Δ: l' nrm̄ *E*. ulla *EGR*: nulla *NV*Δ. etiam si *VR*.

Ep. 93] *nova ep. in EVR; superiori iungunt GN*Δ.

1, 4 δέρρεις *vulg.*: ΔΕΡΡΙC *vel sim. GR*Δ*c*.

6 an *Sternkopf*: nam Σ: *om. Mm*.

7 latinas Δ.

8 advolas et *vulg.*: -asset *N*: -asses et *OR*: -as sed Δ.

9 germane Ω: germa Z^b. putamina? vide *Tucker* (peti vide *Manutius*): putavi de Ω. nummos *OR*: -mis *N*Δ.

folks or for me, any sincerity, any sound sense even or intention of enjoying the good things that are yours, you ought to be on your way and nearly home by now. Believe me, I do not find it easy to be without you—and no wonder I miss *you*, when I miss even Dionysius so keenly. My boy and I will both be demanding him from you when the time comes. The last letter I had from you was dispatched from Ephesus, 9 August.

93 (IV. 19)

Rome, end of November 54

CICERO TO ATTICUS

1 At last the letter I have been longing for! Delighted to hear you are back! I salute you as a man of your word, a paragon of punctuality. And what a lovely voyage! I was really nervous about that I assure you, remembering those tarpaulins on your last crossing. But if I am not mistaken I shall see you sooner than you say, for I imagine you thought your womenfolk were in Apulia. When you find that this is not the case, what is to keep you in Apulia? Or do you have to spare some days for Vestorius and resavour his Latin Atticism after the interval? No, no! Hurry back to Rome, come and look at the empty husks of the real old Roman Republic we used to know. For example, come and see money distributed before the elections tribe by tribe, all in one place openly, see Gabinius acquitted, get the smell of a Dictatorship in your nostrils, enjoy
2 the public holiday and the universal free-for-all, behold my equanimity, my amusement, my contempt for Selicius' 10%, and, yes, my delectable *rapprochement* with Caesar. That does

10 olface *Sternkopf*: cociace (-i ace Z^b) ΔZ^b(*post* nunc; *vide ad* 91. 3, 2): eciate *N*: ocio *V*.
11 dictatum fluere Z^b.
2, 2 Selicianae *C*λ: seleuc- *EN*Δ: seleutice *G*: ceciliane (cic-) *VR*: seleptiane *P*.
3 unciae *C*: provinciae Ω.

haec enim me una ex hoc naufragio tabula delectat. qui quidem Quintum meum tuumque, di boni, quem ad modum tractat honore, dignitate, gratia! non secus ac si ego essem imperator. hibernam legionem eligendi optio delata commodum, ut ad me Quintus scribit. hunc tu non ames? quem igitur istorum?

Sed heus tu, scripseramne tibi me esse legatum Pompeio et extra urbem quidem fore ex Id. Ian.? visum est hoc mihi ad multa quadrare. sed quid plura? coram, opinor, reliqua, ut tu tamen aliquid exspectes. Dionysio plurimam salutem; cui quidem ego non modo servavi sed etiam aedificavi locum. quid quaeris? ad summam laetitiam meam quam ex tuo reditu capio magnus illius adventus cumulus accedit. quo die [ad me] venies, tu, si me amas, apud me cum tuis maneas.

2, 15 accedit Ω: -det *c.* ad me *seclusi* (advenies *coni. Corradus*).
16 tu *Faërnus*: ut *G codd. Mal.* Δ: et *N*: *om. EVR.*

give me some satisfaction, the one plank left from the wreck. The way he treats my (and your) dear Quintus! Such distinction, such appreciation and favour! I couldn't do more if I were G.O.C. myself. He has just offered him the choice of a legion for the winter, so Quintus writes to me. Wouldn't you love such a man? If not, then which of your fine friends?

Now pay attention. Did I tell you that I have joined Pompey's staff and shall be out of town after the Ides of January? I thought this would fit in nicely from several points of view. But I won't ramble on. The rest, I think, when we meet—might as well leave you *something* to wait for! My kindest regards to Dionysius; I have not only kept a place for him, but actually built one. In fact the prospect of his arrival makes no small addition to the great happiness I feel in your return. Do please stay with me your first night in Rome, you and your people.

COMMENTARY

46 (III. 1)

The letter seems to have been written as C. set off on his journey into exile, perhaps from Formiae or Arpinum. On the date and circumstances and on his subsequent movements see Appendix I.

2 **rogationem** *de exsilio Ciceronis*: see Appendix I, p. 227.

4 **consequare** *consequerere* Wesenberg, perhaps rightly; but cf. Sjögren, *Comm. Tull.* p. 126. I cannot find anything in C. quite parallel to the historic tenses after *consequare*, but it is hard to be sure that they go beyond permissible limits, especially in a letter written under stress. The corruption too would not be very easy (though possibly due to *consequare* below).

8 **quod...potes** No *need* for *quod eo.* For *potes* (sc. *me consequi*) Sjögren (*Comm. Tull.* p. 153) compares *Tusc.* I. 83 *quid ego nunc lugeam vitam hominum? vere et iure possum; sed* sqq. But so many of Cratander's supplements have to be accepted that it seems unwise to reject this one.

9 **lex** The law assigning Macedonia to Piso and Cilicia (later changed to Syria: cf. *Dom.* 23) to Gabinius was synchronized with that *de capite civis Romani* (cf. Appendix I) 'to prevent double-crossing' (R. G. M. Nisbet, *Cicero in Pisonem*, Intr. p. x, citing *Sest.* 25; *Pis.* 21). Atticus' business interests in Greece will have made him reluctant to leave Rome before the new governor of Macedonia had been appointed.

47 (III. 3)

See Appendix I, p. 228.

2 **quidem** It is scarcely worth while to contend for *equidem* (cf. K.–S. I, pp. 806 f.).

3 **multis** Besides the explanation in his next letter, C. was thinking of a refuge in Sicily (cf. *Planc.* 95 *Siciliam petivi animo*) or Malta (cf. 49 (III. 4)).

4 **sed** *scilicet* Wesenberg (but the assumed contraction in the archetype is dubious: cf. 22 (II. 2). 2 n. (*sed...doceo*)), del. Kayser. Lehmann detected a contrast between *quo...iter meum* and *de toto itinere*, comparing for *sed... sed* 413 (XVI. 3). 3 *sed exspecto quid ille tecum, quid tu vicissim, nec dubito quin suo more uterque. sed eas litteras Curium mihi spero redditurum.* Here the second *sed* may be simply resumptive, after *quo...meum.*

48 (III. 2)

1–2 **pro meo iure** '*As* of right', *veniam non petens*: see Madvig on *Fin.* V. 75, who cites among other passages *Verr.* II. 5. 2 *tum etiam pro suo iure contendet ne patiamini* sqq. So *iure suo*; cf. *Cael.* 38 *cuius in hortos...iure suo libidines omnium commearent*; Mart. V. 62. 1 *iure tuo nostris maneas licet, hospes,* *in hortis.* *suo iure* in Petr. 78. 7 *cum aqua securibusque tumultuari suo iure coeperunt* does not = *suo more.* C.'s relations with Sicca were such as to let him feel that he could stay as long as he liked: cf. 414 (XVI. 6). 1 *VIIII Kal. igitur ad Siccam. ibi tamquam domi meae scilicet.* This man, often mentioned in letters to Atticus and Terentia and always by cognomen only, had promised to accompany C. into exile, but actually left him at Brundisium. Plutarch (*Cic.* 32) has what seems to be a garbled account according to which Οὐΐβιος, Σικελὸς ἀνήρ (Sicca of Vibo?), who had been C.'s *praefectus fabrum* in 63 and was indebted to him for many favours, offered to make over to him (καταγράψειν) his farm (χωρίον) near Vibo, but did not receive him in his house. See also 420 (XVI. 11). 1 n.

2 **fundum** *in fundo* ς: but cf. with Sjögren 275 (XII. 36). 1 *habeo ne me quidem ipsum quicum tam audacter communicem quam te.*

2–3 **nondum...correcta** C. had heard that Clodius intended to modify his bill (*de exsilio Ciceronis*), but did not yet know what the modification would be (cf. 49 (III. 4)). He needed somewhere to wait and see.

5 **Autronium** P. Autronius L. f. Paetus, C.'s former schoolfellow, friend, and colleague as Quaestor in 75 (*Sull.* 18). He and P. Sulla were elected Consuls in 66, but a conviction *de ambitu* deprived them of the office. Involved in the abortive Pisonian conspiracy of 65 as well as in that of 63, he was condemned *de vi* the following year. C., who refused to undertake his defence, gives a lively sketch of him in *Sull.* 71 and charges him with having incited C. Cornelius to assassinate himself (*ibid.* 18). He was now an exile in Greece. His son obtained a Consulship in 33, and a Triumph.

9 **VI Kal.** See Appendix I.

Narib. Luc. ⟨*ad*⟩ *Naris Luc.* (Boot) may be right, as Nares Lucanae seems to have been originally the name of a mountain pass (cf. Philipp, *RE.* XVI. 1715).

49 (III. 4)

4 **audieramus** Cf. 48 (III. 2) n. *nondum rogatione correcta.*

erat Cf. K.–S. I, p. 123.

4–5 **quadringenta milia** Sc. *ab Italia*: cf. 52 (III. 7). 1 *ab Italia*; Plut. *Cic.* 32 Ἰταλίας. Plutarch says 500 miles, agreeing with Dio, XXXVIII. 17. 7 (3750 stadia). Since it is hardly possible to suppose a slip on C.'s part or an error in his copy of the bill, either *quadringenta* is corrupt or the later authorities mistaken. C.'s uncertainty about Athens (cf. 52 (III. 7). 1 n.) does not prove the former.

5 **liceret...liceret** *esse* and *pervenire* are emphatically contrasted. The bill as amended left C. the privilege of living outside the stated geographical limit but did not allow him means to reach it, since it provided that his 'interdiction from fire and water' should commence from the date of its

passage into law (*diem rogationis*: cf. *Dom.* 47 *non tulit 'ut inderdicatur'. quid ergo? 'ut interdictum sit'*, *ibid.* 82). The original bill probably applied to the whole *imperium Romanum*, but left C. free to settle in any *foederata civitas*, such as Malta (see below). Cf. Sternkopf's lengthy discussions in *Philol.* 59 (1900), pp. 288 ff. and 61 (1902) pp. 42 ff. It is unlikely that C. used the rare, anteclassical form *illoc.*

6 **ne et Sicca** Usually considered a hyperbaton, = *et ne Sicca.* More probably *et* = *etiam* as in 340 (XIII. 37*a*) *habet et Balbus*; 363 (XIV. 9). 3 *et Balbus hic est*; *Q.Fr.* II. 14. 3 *id et Curtius ita volebat.* Harbouring C. within the prescribed limit would involve outlawry under the bill: cf. Dio, XXXVIII. 17. 7 καὶ προσεπεκηρύχθη ἵν' εἰ δή ποτε ἐντὸς αὐτῶν φανείη, καὶ αὐτὸς καὶ οἱ ὑποδεξάμενοι αὐτὸν ἀνατὶ διόλωνται.

7 **quod...non licebat** Malta is less than 200 miles from the Italian coast. Moreover C. had heard at Vibo that his friend C. Vergilius, governor of Sicily, would not give him permission to land (*Planc.* 96). That he says nothing to Atticus about the latter point is not surprising. It would have been painful and unnecessary, since Malta was now ruled out anyway.

11 **haec** *hoc* Wesenberg. *haec* = *id quod scripsi et eius similia*, as often: cf. Lehmann, *Quaest.* pp. 76 ff.

50 (III. 5)

2 **vivo miserrimus** Cf. *Fam.* XIV. 1. 2 *beatissimi viveremus*; Catull. 10. 33 *sed tu insulsa male et molesta vivis.* Often so in Plautus, e.g. *Cas.* 403 *hic vincet, tu vives miser*, *Men.* 908 *ne ego homo vivo miser.*

5–6 **ut...ut** Such repetitions are common: see Sjögren *ad loc.*

8 **VIII** See Appendix I.

Thuri From *Thurium* (Θούριον Diod. II. 12. 6; Ptol. *Geogr.* III. 1. 12, etc.): cf. Plin. *N.H.* III. 97; Mela, II. 68.

51 (III. 6)

2–3 **et ut...et** Cf. 111 (V. 18). 4 *et ut tu...et nos*; Sjögren, *Comm. Tull.* pp. 137 f.

2 **consisteremus** Cf. 46 (III. 1).

7 **Tarentino** A territorial neuter: cf. 23 (II. 3). 3 n. (*Solonium*).

52 (III. 7)

1, 1 **Brundisium** C. stayed there thirteen days in the *horti* of M. Laenius Flaccus: cf. *Fam.* XIV. 4. 2; *Planc.* 97.

5 **esset** It is possible to retain *sed*, reading *sic itineris* below. The anacoluthon would not be strongly objectionable in itself, but I hardly think that C. started his sentence with a phrase which was obviously going to land

him in syntactical complications. *sed ut...sic* (Le Clerc) is a respectable alternative. W. S. Watt (*Eranos*, 60 (1962), p. 68) proposes *scilicet*, but cf. 22 (II. 2). 2 n. (*sed...doceo*).

7 **lucem** Cf. 64 (III. 19). 1 *qui lucem omnino fugerem*. The association of *lux* with publicity, 'the sight of men' (cf. 108 (v. 15). 1), makes such expressions the more natural.

10 **quadridui** Sc. *viam abest* (cf. Liv. XXV. 8. 12 *tridui viam aberat* et sim.).

nam 'Occupatory.' Atticus had no doubt pointed to the strength of his Buthrotian 'castle' (*castellum munitum* may be quoted from his letter) as one reason for C.'s going there.

12 **cadebat** sc. *res*: cf. *Brut.* 149 *sed ita cadebat ut alter ex alterius laude partem, uterque autem suam totam haberet*; *Fam.* II. 19. 1 *verebar ne ita caderet, quod etiam nunc vereor, ne...decederem*. The point I take to be that before his exile C. might well have *wanted* to retire to Athens to escape conditions in Rome. As a *legatus* he would have had less or nothing to fear from his enemies there. Now that plan is impossible.

14 **illud quoque oppidum** As well as Buthrotum. For *oppidum* applied to Athens cf. 21 (II. 1). 2 *ut et Athenis sit et in ceteris oppidis Graeciae*; *Fin.* v. 96 *in oppidum*. The doubt about the distance seems to have lain in whether calculation would be based on the route Brundisium—Dyrrachium—Athens or on the route Brundisium (or Hydruntum)—Corcyra—Actium—Athens. The former gives well over 500, the latter something under 400 miles: see Sternkopf, *Philol.* 59 (1900), pp. 292 ff. Thessalonica, where C. actually settled, was reckoned as 392 miles from Brundisium.

2, 3 **spes** Cf. *Q.Fr.* I. 4. 4 *haec mihi proficiscenti non proponebantur, sed saepe triduo cum summa gloria dicebar esse rediturus*.

5–6 **per...scelus** I.e. through their advice to withdraw instead of facing the music after Clodius had promulgated his law *de capite civis Romani*: cf. 54 (III. 9). 2 *nos non inimici sed invidi perdiderunt*. C. often uses this sort of language (cf. 53 (III. 8). 4 n.). He certainly had Hortensius in mind, perhaps Arrius (cf. *Q.Fr.* I. 3. 8), not Cato (cf. 60 (III. 15). 2). It seems unlikely that his charges were well founded.

10 **honestissimum** 'Cum inimicis pro mea dignitate pugnanti' (Manutius): cf. *Q.Fr.* I. 3. 6 *scio fuisse et honestius moriendi tempus et utilius*.

11 **non iam sqq.** The moment for an honourable death having gone by, C. must live; but from his remaining time he looks for no healing, only for, eventually, an end to his troubles. The sense is not expressed with perfect clarity (*tam...quam* makes it rather easier), but intelligibly enough. One is reminded of Admetus' reply in Euripides, when he was told that time would soften his grief: χρόνον λέγοις ἄν, εἰ χρόνος τὸ κατθανεῖν (*Alc.* 1086). *esse ad* has various nuances: cf. *Nat. Deor.* III. 83 *laneum pallium iniecit, cum id esse ad omne anni tempus diceret*; Ter. *Heaut.* 207 *haec sunt tamen*

ad virtutem omnia; Caes. *B.C.* III. 101. 2 *reliquisque rebus quae sunt ad incendia*; Vitr. x. 7. 5 *reliqua quae non sunt ad necessitatem sed ad deliciarum voluptatem* et sim.

11 **medicinam** Cf. *Fam.* v. 16. 6 *exspectare temporis medicinam.*

3, 5 **Candaviam** District of Macedonia, north of Epirus.

8 **quo modo** Cf. 54 (III. 9). 1 (especially *etiam illud timebam...ne a me digredi non posset*); 55 (III. 10). 2; *Q.Fr.* I. 3. 4. For the variation *quo modo... ut* cf. *Rep.* II. 43 *quamvis...ut*; *Phil.* v. 14 *quem ad modum...quo modo*; Tac. *Hist.* III. 77 *nec virtutibus, ut boni, sed, quo modo pessimus quisque, vitiis valebat.*

9 **id** Generally, the thought of Quintus in connection with C.'s present plight.

13–14 **d....proficiscens** Cf. 176 (IX. 9). 4 *d. Liberalibus.* As may easily have happened elsewhere *d.* = *dedi* or *dabam* was transcribed *data.* It is better (though not absolutely necessary) to read *Brundisio* (Klotz), than to transfer *proficiscens* to the following letter (see below). Since C. wrote as he was leaving he would naturally say so: cf. 359 (XIV. 5). 3 *haec scripsi ad te proficiscens Astura III Id.*; *Fam.* XVI. 5. 2 *Leucade proficiscens VII Id. Nov.*

53 (III. 8)

1, 1 **scripseram** Even if 52 (III. 7) is the letter in question, I see no advantage in reading ⟨*Brundisio*⟩ *proficiscens* (Wesenberg: see above) before *scripseram.* But this seems doubtful. In 52 (III. 7) C. was still thinking of going to Epirus (§ 3). Here he says he gave two reasons for deciding not to do so of which the second is not mentioned in 52 (III. 7). 1. He may well have written again after landing at Dyrrachium.

3 **proficisceremur** For the tense cf. 54 (III. 9). 1 *cum ille...dimitteret*; 183 (IX. 15). 3 *cum sciremus*; Liv. XXII. 43. 11 *salutare praecipue futurum erat cum aciem dirigerent.*

2, 1 **quid agatur...timeo** Cf. K.–S. II, p. 487.

2 **postularetur** Sc. *frater.* There is a marginal possibility that *postularet* (sc. *Clodius*) is right, but the change is very easy.

3 **esse mitiora** Such expressions are not uncommon in C.'s letters, as 124 (VII. 1). 9 *puta* ἄμεμπτα *fore*; 326 (XIII. 19). 1 *tu* ἀκίνδυνα *esse scribis*; 368 (XIV. 14). 7 *posse videntur esse meliora*; *ad Brut.* 18. 5 *velim deinceps meliora sint.*

8 **Phaëtho** Cf. *Q.Fr.* I. 4. 4 (early August) *in ea epistula quam Phaëthonti dedi.*

9 **Ilio** *illo* (i.e. *Quinto*) is patently impossible. *Ilio* would be equally so if this freedman were the messenger to Athens mentioned in § 1, but in that case C. would probably have named him in the first place. Phaëthon will have been sent to Asia (as Atticus apparently knew) before C. left Italy.

12 **nostram** 'Mine and my brother's', as *equidem* indicates.

3, 1 **Tryphonem** Probably a freedman of Atticus' uncle Caecilius.

4 **Tigrane** Son of the King of Armenia, taken to Rome by Pompey and produced at his Triumph, then handed over to L. Flavius (cf. 18 (I. 18). 6 n.) for safe-keeping. His liberation by Clodius had resulted in a scuffle on the Appian Way in which lives were lost (Ascon. 47. 12). For *Tigrane* = *Tigranis negotio* cf. 280 (XII. 39). 2 *nisi Otho exstitisset*; 285 (XII. 44). 2 *iam inde a Pontiano*; 298 (XIII. 27). 2 *operam in Faberio ponamus*; 300 (XIII. 29). 2 *si Faberium explicas*; 356 (XIV. 2). 1 *theatrum Publiliumque cognovi*; *Fin.* II. 44 *cum Epicuro autem hoc plus est negoti quod e duplici genere voluptatis coniunctus est.*

6 **Hypsaeo** P. Plautius Hypsaeus, of consular family. A leading adherent and former Quaestor of Pompey, he was now Curule Aedile, later Praetor, and a candidate for the Consulship in 53. Pompey allowed him to be condemned *de ambitu* in the following year.

acta Probably a copy of the *acta urbana* (cf. 116 (VI. 2). 6), the official record of state proceedings first regularly published in 59 (cf. Suet. *Iul.* 20. 1). Cf., however, 55 (III. 10). 1 *acta quae essent usque ad VIII Kal. Iun. cognovi ex tuis litteris*; 67 (III. 22). 1 *quae essent acta scripserant.*

8 **ubi** 'Where I shall stay (when I do leave)': cf. 55 (III. 10). 1 *quibus adlatis facilius statuere potuero ubi sim.* Not 'where I shall stay in the meantime'.

4, 2–4 **motum...commotus** Cf. 217 (XI. 6). 3 *quo magis factum nostrum Caesar probet quasi de suorum sententia factum.* Both letters were written *de profundis*, but cf. also *Verr.* II. 3. 22 *aspicite, iudices, vultum hominis et aspectum*; *Rab. Post.* 40 *sermo...auris refersit istis sermonibus*; Trebonius, *Fam.* XII. 16. 3 *dictum...a te dictum.*

4 **cuius** Hortensius is meant: cf. 54 (III. 9). 2, *Q.Fr.* I. 3. 8, also 52 (III. 7). 2 *invidorum*; 55 (III. 10). 2 *nisi intra parietes meos de mea pernicie consilia inirentur*; 59 (III. 13). 2 *qui, ut me paulum inclinari timore viderunt, sic impulerunt ut omni suo scelere et perfidia abuterentur ad exitium meum.* The last passage fixes the point of *impulsi.*

6 **maerori** *errori* Madvig. But *impulsi*, taken in connection with 59 (III. 13). 2, shows that C. has in mind his now regretted decision to leave Rome after the promulgation of Clodius' bill *de capite civis Romani.* There was reason enough for *maeror* then: cf. 60 (III. 15). 4 *tu tantum lacrimas praebuisti dolori meo* (there too he complains that Atticus had not *advised* him). No doubt *errori* would pass if it were in the MSS (cf. 73 (IV. 1). 1 *te qui primis temporibus erroris nostri aut potius furoris particeps et falsi timoris socius fuisses*), though *totum animum dedisses* suits *maerori* better.

8 **eventi** *poenam eventi* is an odd phrase: cf. *de Orat.* I. 202 *iudiciorum poena*, 'penalty inflicted by the courts'. But no doubt C. really means

eventum (conjectured by Wesenberg), and was seduced into writing *eventi* to make a false parallel with *stultitiae.*

54 (III. 9)

1, 3 **calamitatis** Quintus was threatened with a prosecution in respect of his government of Asia: cf. §3 and 62 (III. 17). 1.

8 **mollissimo** Cf. 17 (I. 17). 2 *quam mollis animus.*

9 **adflictus** Reid compares the parallel passage in the next letter (§2) and *Phil.* IX. 12 *adflictus luctu. miserias luctu adflictas* is surely a phrase that C. could never have written in any state of mind.

12 **lictores dimitteret** Quintus could not remain in Greece indefinitely while retaining his *imperium*, though returning governors could travel home very much at their leisure. C. himself took nearly four months over his return from Cilicia to Italy in 50.

15 **peccati** Cf. *Fam.* XIV. 4. 5 (to Terentia) *peccatum est nullum nisi quod non una animam cum ornamentis amisimus.*

2, 1 **quantum** Not quite equivalent to *quantulum*. There is always an ironic flavour about *quantus* used thus.

8 **si...speras** Lit. 'If the things you hope for are so', i.e. 'as you represent them': cf. *Acad.* I. 11 *si haec ita non sunt* et sim. *ista* (Schütz) is not an improvement.

9 **infirma sunt** Sc. *quae speras. firma* can hardly stand, not so much because of the difficulty of supplying a subject from *ut mihi videntur* (Plaut. *Asin.* 566 *fateor profecto ut praedicas, Leonida, esse vera* would be rather similar) as because of the oddity of the word; we should expect *vera.*

9–10 **quod...non licuit** I.e. death, whether by his own hand or in conflict with Clodius (cf. 52 (III. 7). 2 n. (*honestissimum*)). If *genere* were retained as equivalent to *modo*, *optimo* would denote the latter and *minus idoneo* the former. But editors are probably right to prefer *tempore* despite the MSS and Cratander's and Lambinus' margins.

3, 1 **etiam** I think this must be taken with *mihi*, as a reference to something in Atticus' letter.

3 **beneficiorum** Perhaps connected with Varro and Hypsaeus (cf. 53 (III. 8). 3). *istorum* may, of course, be pronoun, not adjective. *comitiorum* (Graevius) is worse than needless. The elections did not take place till July, whereas Atticus' advice concerned events in May (*ibid.*; cf. 55 (III. 10). 1).

4 **litterarum exspectatio** Cf. 56 (III. 11). 1 *exspectatio vestrarum litterarum.*

5 **si quid** Practically = *simul atque aliquid*: cf. 30 (II. 12). 3 n.

6 **ut scribis** Atticus will have written *Kal. Iun. profecturus* (or *profectus*) *sum*; but something must have happened to change his plans, for he did not leave Italy till the following year.

55 (III. 10)

1, 3 causa Not 'reason' or 'excuse'. As in 39 (II. 19). 2, 126 (VII. 3). 5, and 153 (VIII. 3). 4 the word implies a cause actively prosecuted, an organized movement.

4 **ad te** I.e. *Buthrotum*. Atticus, having supposedly left Rome on 1 June, would be there to receive him, though the text does not necessarily so imply.

5 **ista** Cf. *Fam.* VII. 11. 2 *sin autem ista sunt inaniora.*

ut tu scribis As Manutius pointed out, Atticus may have written something like *si haec evanuerint, alia videbimus* without expressing the opinion that hopes would indeed prove dupes: cf. 150 (VII. 26). 3 *nisi qui, ut tu scribis, Parthicus casus exstiterit*; 271 (XII. 32). 2 n. *ages, ut scribis, temperate.* There is therefore no need to accept the transposition (an old one, known to Corradus) *me ad te conferam, ut tu scribis.* On the other hand, T.–P.'s contention that 'Atticus did *not* write him reassuring letters' is wholly unacceptable. Of the two passages they cite in support of it, 58 (III. 14). 1 and 59 (III. 13). 1, the first is corrupt and the second makes if anything the other way; whereas the correspondence abounds in evidence that around this date Atticus had at any rate tried to sound hopeful. 29 April (52 (III. 7). 3): *de re publica video te colligere omnia quae putes aliquam spem mihi posse adferre mutandarum rerum.* 13 June (54 (III. 9). 2): C. says he is sustained and comforted by Atticus' letter, though he could see between the lines that his friend was not really hopeful. 27 June (56 (III. 11). 1): C. is stopping at Thessalonica because of Atticus' letters and *quidam boni nuntii.* 17 July (57 (III. 12). 1): *tu quidem sedulo argumentaris quid sit sperandum...spem ostendis secundum comitia.* 5 August (59 (III. 13). 1): Atticus, having nothing cheerful to write, had not written. August, first half (*Q.Fr.* I. 4. 2): C. had stayed in Thessalonica on the strength of hopeful reports from Atticus and others. 17 August (60 (III. 15). 6): Atticus has been hiding unpalatable news. 19 August (61 (III. 16)): a plea for greater frankness; the more C. studies his friend's letters the plainer it appears that though hopeful in tone they contain no solid ground for encouragement. Atticus does now seem to have written in a less optimistic vein. 4 September (62 (III. 17). 2): *cetera quae ad me eisdem litteris scribis de nostra spe intellego esse languidiora quam alii ostendunt.* First half of September (63 (III. 18). 2): *tuae autem litterae sunt variae; neque enim me desperare vis nec temere sperare.* 15 September (64 (III. 19). 2): *me tuae litterae numquam in tantam spem adduxerunt quantam aliorum. ac tamen mea spes etiam tenuior semper fuit quam tuae litterae*, i.e. Atticus had been less encouraging than, for example, Quintus, who *omnia mittit spei plena* (63 (III. 18). 2). 10 December (69 (III. 24). 2): *tota res quo loco sit velim ad me scribas et ita ut instituisti. nam ista veritas, etiam si iucunda non est, mihi tamen grata est.*

Here therefore translations like 'si en revanche, comme tu me l'écris, ces chances-là doivent s'évanouir' are ruled out by the facts of the case. Even if they were not, they would be discountenanced by what follows. The argument that division among the 'Triumvirs' gave no solid ground for hope implies that Atticus had suggested that it did; and Atticus is included in *vos me sperare vultis*.

5 **aliquid aliud** I follow λ here though *alius aliquis* is the normal Ciceronian order: cf. Sjögren, *Comm. Tull.* p. 84. Varro and Lucretius sometimes have *aliquis alius*. Perhaps there is a slight difference of emphasis, 'something other (different)' as distinct from 'some other thing'.

6 **omnino** Concessive, taken up by *sed tamen*.

significatis Sc. *ad spem*.

9 **vos** Atticus and other well-wishers. Terentia and Piso may be included, and perhaps Quintus, who returned to Rome about the beginning of June.

2, 3 **tam ex amplo** *tam extemplo* ⟨*tam ex amplo*⟩ Persson, followed by Sjögren. But *tam extemplo* = *tam subito* is hardly C.'s Latin. Indeed *extemplo* is found only once in his prose, in an early speech (*Rosc. Com.* 8 *quid fingat extemplo non habet*).

6 **qui fuerim** This seems to be the classical expression: cf. 60 (III. 15). 8 *qui fui et qui esse potui, iam esse non possum*; *Fam.* VII. 3. 4 *vetus est enim, ubi non sis qui fueris, non esse cur velis vivere*; *Lig.* 7 *ex Aegypto litteras misit ut essem idem qui fuissem*; Prop. I. 12. 11 *non sum ego qui fueram*. In Ovid, *Tr.* III. 11. 25, where the MSS vary, editors read *non sum ego quod fueram*: cf. *Propertiana*, p. 38.

14 **aut non** *non aut* Madvig, removing a very awkward inversion. But it may be that C. himself did not quite see his way through his own labyrinth. His meaning is plain enough.

15 **inirentur** By Hortensius and other counsellors of retreat: cf. 53 (III. 8). 4 n. (*cuius*).

3, 3 **quod et...et** C. might have written *et quod...et quod* but for the clumsiness of *et quod quod exspectem*, but it is as far wrong to speak of an ellipse of *quod* here, let alone of a very harsh ellipse, as it would be in e.g. 64 (III. 19). 1 *quod et celebritas mihi odio est et...abesse longe nolebam* or 189 (IX. 19). 4 *quod et...tenui et...perseveravi*. Such an ellipse does occur in *de Orat.* II. 246 *est hoc scurrile et quod sine causa lacessivit et tamen id dixit quod in omnis luscos conveniret*, unless *et quod* is there to be regarded as an inversion of *quod et*.

56 (III. 11)

1, 3 **si accepero** Quasi-temporal: cf. 30 (II. 12). 3 n.

4 **ad te** Cf. 55 (III. 10). 1 n.

2, 2–3 **quod...desidero** This seems to soften the preceding prohibitions. Even Atticus' scolding revealed his affection and concern, and so made C. long for his presence. Or possibly *quod* refers only to *consolari*, *obiurgare vero noli* being parenthetic. The words are usually understood with Manutius: 'facis ut neque me videaris amare neque dolere dolorem meum.'

6 **omnia certa** Cf. *Fam.* XVI. 24. 2 *de publicis omnia mihi certa* (sc. *scribe*); 424 (XVI. 13 *a*). 1 *omnia igitur velim explices et ad me certa mittas*, 2 *tu mihi...omnia certa, clara*; 114 (V. 11). 14 *velim ad me scribas certum quo die mysteria futura sint*; Liv. XXXVII. 39. 5 *postquam omnia certa rettulit* et sim.

57 (III. 12)

1, 2 **rogationis** Clodius' bill *de exsilio Ciceronis.*

proponi I.e. *in curiae poste* (60 (III. 15). 6).

6 **eodem** Clodius would remain in office until 10 December. Editors, perhaps misled by Manutius, imagine that C. expected him to be re-elected.

inimico Metellus Nepos. In the event he declared on entering office (and privately some two months earlier: cf. 69 (III. 24). 2) that he would not oppose C.'s restoration, with what sincerity is doubtful (cf. Münzer, *RE.* III. 1218. 1): cf. C.'s letter to him, *Fam.* V. 4.

2, 2 **oratione** *in Clodium et Curionem*, composed some three years previously in the aftermath of the Bona Dea affair. Quintilian (*Inst.* III. 7. 2) refers to it as extant, and fragments survive. It was partly based on the speech and *altercatio* in the Senate recounted in 16 (I. 16). 8 ff.

3 **medere** Probably Atticus' word, though C. is fond of the metaphor.

ille Curio the Elder: cf. 14 (I. 14). 5 n.

6–7 **quia...ceterae** Not of course put forward as a *reason* for denying authenticity, but as a feature that would make the denial more credible.

7 **ex se posse** 'In structura *ex se probari* ne quis haereat, cfr. similes has: de Fin. II: 26[83] *quae faciat amicitiam...ex se et propter se expetendam*; de N.D. II: 12[32], III: 14[36]' (Sjögren, *Comm. Tull.* p. 34). Add, more relevantly, Caes. *B.G.* V. 27. 4 *id se facile ex humilitate sua probare posse.* Without *posse* the sense would be 'I think that it is being' (or 'that it is going to be') 'successfully passed off', etc.: cf. *ad Brut.* 5. 4 *hoc cogere volebat falsas litteras esse et, si quaeris, probabat*; *Verr.* II. 5. 78 *in eius locum quem pro illo probare velles*; *Rab. Post.* 41 *vel, si meminerit, oblitum esse ⟨se⟩ facile possit probare*; *Mil.* 65 *mirabar ...vulnus in latere, quod acu punctum videretur, pro ictu gladiatoris probari.*

3, 1 **iaceo** Cf. 217 (XI. 6). 2 *Brundisi iacere.*

2 **licet...significarim** *scilicet...significaram* Madvig, because *licet* = 'although' is un-Ciceronian (cf. K.–S. II, p. 444); but see my translation.

3 **id dono tamen et** Springer (*si non, dono tamen et*) and Constans (*condono tamen et*) saw the essentials. For *dono* = *condono* thus cf. *Thes.* V (i).

2015. 52, 2014. 53. It is true that the examples are post-Augustan, but the same objection might be made to *tuas inimicitias ut rei publicae donares* in *Fam.* V. 4. 2 if the MSS there had happened to be corrupt (cf. *Thes.* v (i). 2014. 46). M. Brutus and Caesar use *donare* of remitting a debt (*ibid.* 2015. 20).

5 **magis** Not = *eo magis*: cf. 55 (III. 10). 3 *quod exspectam istinc magis habeo quam quod ipse scribam.*

58 (III. 14)

1, 3 **agi** Sc. *in senatu* (Manutius).

4 **et scio sqq.** The vulgate *et* (or *etsi*) *scio te me iis epistulis* sqq. is supposed to mean 'and I mind me that it is rather your wont in your correspondence with me to check and discourage me and my hopes' (T.-P.). Such a sense would be at odds with the evidence of other letters (cf. 55 (III. 10). 1 n.); and the Latin is gibberish. Even when *iis* is ameliorated with Purser to *istis* or *tuis*, there remains the futility of *me et meas spes*, emphasized by a distracted word-order. Furthermore the meaning attached to *spes remorari* is dubious. The analogy of Ter. *Andr.* 739 *ne quod vestrum remorer commodum* would rather suggest 'put off the fulfilment of hopes'. But perhaps, as Manutius and others thought, *remorari* means 'keep in being': cf. *ad Herenn.* IV. 48 *et iam eum cuius opera nobis haec accidunt vos remoramini diutius et alitis ad rei publicae perniciem, retinetis, quoad potestis, in civitate?* A verb is then required before *potius*, so that rewriting might take the form *et scis te me istis epistulis* ⟨*sustentare*⟩ *potius* sqq.

2, 3 **Plancio** Cf. 12 (I. 12). 2 n., also 67 (III. 22). 1 and *Planc.* 100 *vi me, vi, inquam, Plancius et complexu suo retinuit.* The governor of Macedonia, L. Appuleius Saturninus, though a *civis optimus* and a friend of C.'s, *eadem timuit quae ceteri* (*ibid.* 99); but Plancius, as Quaestor, was able and willing to give him protection. Since Thessalonica was within the prohibited distance from Italy (cf. 49 (III. 4) n.) he ran some risk (cf. *Planc.* 99), in theory at least. In practice C. does not now seem to have troubled himself about this regulation; perhaps he had received some assurance that it would not be pressed.

5 **calamitatem** Often with reference to the loss of civic rights (*salus*).

ut scripseram Cf. 59 (III. 13). 1. Unless he refers to a missing letter, C.'s memory is not quite accurate. He had written that he might or would go to Epirus *if* he received hopeful news: cf. 55 (III. 10). 1; 56 (III. 11). 1.

veni Perhaps epistolary, 'am not going'. C. rarely uses the perfect of *ire* and never in the first person; he often uses *veni* instead (cf. Sjögren, *Eranos*, 16 (1916), pp. 41 f.).

6 **universi** *adversi* (Koch) or *non varii* (Watt) makes the expression more logical; but the continuation *qua re...esset* can be regarded as a natural if loose alternative to *eius modi ut...esset.*

7–8 **si...audierimus** *nisi* Manutius. But *si* is virtually temporal: cf. 30 (II. 12). 3 n. After recent depressing reports C. assumes that when the news comes it will be unfavourable.

59 (III. 13)

1, 8–9 **meque...non moleste** The alternative to *non* is *neque me* (Ernesti) for *meque*. Perhaps it is not absolutely certain that any change is needed. 'I shall reproach myself for having been beguiled by hope for a short period' is not nonsense; but it does sound unduly petulant. Supplying the negative, editors take C. to mean that he is glad to have the period of hope so soon cut short; that seems to me to follow less naturally from his words than my rendering. In any case *temporis non longinqui spe* is literally 'by a hope of (i.e. directed towards) a not remote point of time'.

9 **motum** Cf. 53 (III. 8). 3.

11 **tribunis pl. designatis** C. could count on a number of these, especially P. Sestius: cf. *Q.Fr.* I. 4. 3 and *Cl. Rev.* N.S. 12 (1962), p. 195.

2, 2–3 **ut neminem...audieris** *ut* = *quo modo*: cf. 57 (III. 12). 1 *cum ita sim adflictus ut nemo umquam.* The subjunctives *videris* and *audieris* are probably assimilated to *videas*, possibly potential, 'you could never see or hear': cf. W. Blase, *Philol.* 63 (1904), p. 637. Wesenberg added *magis* after *audieris*, comparing the common *nihil vidi minus.*

10 **nunc** *nunc* returns from the past to the present. *iam* = 'presently'.

11 **perferentur** Sjögren's defence of the present (*Eranos*, 16 (1916), p. 40) as 'generaliter dictum' is belied by *ad me.*

60 (III. 15)

2, 3 **ubi...quibuscum** Cf. 57 (III. 12). 3 *sine sermone ullo, sine cogitatione ulla.*

7 **scindam** *rescindam* (Graevius) is not an improvement. C.'s wounds had never closed.

13–14 **me ipsum** Cf. 155 (VIII. 7). 2 *memet ipso carendum est*; *Fam.* III. 10. 10 *memet ipsum mihi per illum restitutum puto*; *post Red.* 1 *nosmet ipsos nobis reddidistis* (sim. *post Red. ad Quir.* 8); *ibid.* 8 *si me mihi, si meis, si vobis, si rei publicae reddidisset*; *Marc.* 13 *me et mihi et item rei publicae...reddidit*; Sen. *Ep.* 104. 6 *repitivi ergo iam me.*

14 **quid enim sum?** Cf. 152 (VIII. 2). 4 *quid ero?*; 249 (XII. 11) *ipsi enim quid sumus...?*; 318 (XIII. 10). 1 *quid enim sumus aut quid esse possumus?*; *Q.Fr.* III. 4. 2 *quid essem, si me agente esset elapsus?*; *Phil.* II. 75 *tu vero quid es?*

16–17 **eos...Catonem** Cf. 52 (III. 7). 2 n., Intr. p. 21, n. 4.

3, 3 **sed quid Curio?** Cf. 30 (II. 12). 2 n. (*quid Caesar?*). No particular verb seems to be understood. C. wonders why Curio should have spoken as warmly as Atticus represented after the appearance of the *in Clodium et Curionem* (cf. 57 (III. 12). 2).

4 **Axius** Cf. 12 (I. 12). 1 n.

5 **non ita** Possibly = *non ita valde* as in Tac. *Hist.* II. 32 *Hispaniam armis non ita redundare.* More probably, 'not as you do'.

at Preferable to *ac* where the sense is *simply* adversative: cf. 44 (II. 24). 1 n.

7 **Varronis sermo** Cf. 63 (III. 18). 1.

4, 1 **ego sqq.** Cf. Intr. pp. 19f., and 177 (IX. 10). 2 where self-reproach makes a platform for condemnation of Pompey. The seemingly questionable taste of the opening promise belongs to a conventional pattern: cf. 372 (XIV. 19). 2 *Antonius ad me...rescripsit, meam lenitatem et clementiam et sibi esse gratam et mihi voluptati magnae fore*; *Fam.* III. 2. 2 *magnam te ex eo et perpetuam voluptatem esse capturum* (*eo* being services to be rendered), XIII. 50. 2 *spondeo...ex mea amicitia et ex tuo in me officio maximum te fructum summamque voluptatem esse capturum* et sim.

7–8 **non...sciam** As the second of two negatives cancelling one another *non* is usually separated from its verb: cf. Hofmann–Sternkopf *ad loc.*

9 **amare** Madvig's denunciation of this word (*Adversaria Critica*, III, p. 170) shows a great critic at his least percipient.

11 **legem** One of four carried by Clodius at the start of his term as Tribune, reviving *collegia* abolished in 64: cf. Dio, XXXVIII. 13. 2; R. G. M. Nisbet on *Pis.* 9. How this measure should have been represented as advantageous to C. or how the delusion injured him is beyond knowledge. Atticus had apparently shared it and might be pricked by the reminder.

12 **quod erat amoris** Cf. 331 (XIII. 23). 3 *fuit id quidem humanitatis* et sim.

13 **potuit** *oportuit* Orelli, needlessly; impersonal *posse* is not rare in these letters. Old editors add *esse*: cf. Varro, *L.L.* V. 118 *quod a nobis media et a Graecis mesa, mensa dicta* (*dici* codd.) *potest*, VI. 58 *quod a verbo Graeco potest declinatum*, 68 *id a* θριάμβῳ...*potest dictum.*

16 **Pompei...responso** Given to a deputation of Senators which visited him at his Alban villa to plead on C.'s behalf (cf. *Pis.* 77). Pompey told them that as a private individual he could not interfere with a Tribune, though he would take up arms if the Consuls invoked the *senatus consultum ultimum.*

18 **aut...occubuissemus** Bruni wrote in M *aut occubuissem honeste*, on which editors have hitherto seen no reason to improve.

20 **me alterum** Cf. 73 (IV. 1). 7 *ad omnia me alterum se fore dixit*; *Fam.* II. 15. 4 *quoniam alterum me reliquissem*; *ad Brut.* 23. 2 *ad te tamquam ad alterum me proficiscens*; *Amic.* 80 *est enim* (sc. *verus amicus*) *is qui est alter idem.*

21 **ac si** *ac* may be the product of the missing *-RO*, but it would be a pity to strike it out.

22 **tuo ipsius** Cf. 402 (XV. 14). 3 *beneficia autem nostra tueri solemus.*

5, 1 **Culleone** Q. Terentius Culleo, of old praetorian family, was Tribune this year. Plutarch (*Pomp.* 49) mentions him as urging Pompey about this time to divorce Julia and break with Caesar. On his earlier conveyance of Atticus' advice to C. see Intr. pp. 17 f. Tyrrell, perhaps rightly, proposed *tecum Culleonem* on the ground that Atticus would not have mentioned Culleo at all if the suggestion had not come from him.

privilegio Laws directed against an individual were forbidden in the Twelve Tables (*Leg.* III. 11). The idea which Atticus had discussed with Culleo was to have Clodius' law *de exsilio Ciceronis* declared invalid on this ground. That would only require a *senatus consultum.* Such a decree was actually proposed by L. Cotta in the following January, but Pompey took the view that a law should be passed as well (*Sest.* 73 f.).

5 **prior lex** *de capite civis Romani.*

8 **sed etiam** Cf. K.–S. II, p. 66.

10 **pergo praeterita** So *pergere reliqua* (86 (IV. 11). 1), *cetera* (*Leg.* II. 69), *maledicta* (*Phil.* XIII. 40). An infinitive (*dicere*) is probably to be understood.

6, 4 **at tute** Cf. 36 (I. 16). 17 n. Probably the archetype had *astute.*

5 **in curiae poste** Cf. 57 (III. 12). 1.

6 **Domitius** Praetor this year and as such constitutionally possessing the *ius referendi*: cf. Mommsen, *St.* I, p. 209. Normally, however, a Praetor (Urbanus) only exercised this right in the absence of Consuls.

10 **domo** After the confiscation of C.'s property under Clodius' second bill his house on the Palatine and his villas at Tusculum and Formiae were looted and destroyed. On part of the site of the house Clodius set up a shrine of Liberty (*Dom.* 100, 110–12); the rest he used for a portico to replace that of Catulus (cf. 74 (IV. 2). 3 n.) and for the extension of his own house (*Dom.* 116: cf. K. Ziegler, *RE.* XVIII. 44. 66).

13 **acta** Cf. 53 (III. 8). 3 n.

16 **id quod** Like Müller I should have kept *idque* 'nisi in codicibus nihil fere interesset inter *que* et *quod* et sim.'

7, 1 **Pomponi** The omission of *mi* is appropriate to a solemn appeal: cf. Caelius, *Fam.* VIII. 16. 1, 3 *Cicero. T. Pomponi* in 64 (III. 19). 3 is even more portentous.

4 **proditus sqq.** Cf. 80 (IV. 5). 1 *inductus, relictus, proiectus ab iis*; 64 (III. 19). 3 *a meis consiliariis proditum et proiectum.*

5 **erectam** Cf. *Sest.* 36 *tota denique Italia ad omnem contentionem expedita. in me*, which may have arisen out of *-m e-* or from a duplication of the following *ad me*, is better deleted than replaced by *iam* (Orelli).

6 **me, meos** Cf. 64 (III. 19). 3 *ut ipse me et meos perderem.* A clumsy

and otiose *meis* (Lambinus) follows *meos* in some texts. Lehmann conjectured *me, meos, mea,* but λ's reading is probably nothing more than a common type of dittography.

8 **adflictos** Sc. *nos,* unless the plural be allusive as in Prop. IV. 9. 34 *pandite defessis hospita fana viris* (cf. Jebb on Soph. *Oed. Col.* 148).

10 **communiter** Cf. 176 (IX. 9). 3 *age, esto; hoc commune est*; Juv. IX. 124 *utile consilium modo, sed commune dedisti* (sometimes misunderstood).

8, 2 **aliquem** Not, I think, quite as Boot explains: 'ita in patriam revocari ut non paeniteat vivere, ut aliqua dignitate sim' (cf. Pers. I. 129 *sese aliquem credens*; Mart. X. 27. 4 *nemo tamen natum te, Diodore, putat*). C. implies that an exile without civic rights is virtually a dead man, has no real existence.

5 **meo nomine** So 212 (XI. 2). 4 *quibus tibi videbitur velim des litteras meo nomine.* Such letters would obviously have to be dictated to a secretary but would usually bear the supposed writer's seal (cf. *ibid.*). Plancus when governor of Asia had permission to use Antony's name and seal in letters (App. *B.C.* V. 144).

61 (III. 16)

3 **aliud quid** *aliud aliquid* Wesenberg: see app. crit.

5 **diligentius** This seems the most natural supplement; others can be and have been proposed. *laetae* would require *quamquam...tamen* or *ita... ut* rather than *cum...tum.*

8 **ut putabis** Usually placed after *quae putabis* for an obvious reason, which does not however outweigh the evidence of Z^bλ.

XII Kal. As C. had written only two days previously he may not have bothered to add the month. The letter could even have been sent along with 60 (III. 15) as a sort of postscript. A scribe is more likely to have added the wrong month than changed *Sept.* (ϛ) to *April.*

62 (III. 17)

1, 2 IIII This explains the lacuna slightly better than *III* (Hand): *d. IIII* may have been read as *di(e)m.*

3 **Livineius** L. Livineius Trypho, of whose kind offices at this time C. later wrote warmly in a letter of recommendation (*Fam.* XIII. 60). The same letter refers to the *calamitas* of his patron, L. Livineius Regulus (*familiarissimi mei*), wherefore Münzer supposes that Regulus was a fellow-exile in Thessalonica. But its wording rather suggests that the *calamitas* came subsequently. Moreover, if Regulus was in Thessalonica, Trypho's journeyings would have been on his account rather than C.'s (cf. *ibid.* 2 *quod pro salute mea multa pericula adierit, saepe hieme summa navigarit*). Regulus held the

Praetorship at some time, so coins of his homonymous son testify (cf. Broughton, *Suppl.* p. 35).

4 **omnino** Concessive, rather than, as usually taken, with *nullam*. The answer is *sed postridie* sqq. rather than *sed fuisse tamen* sqq.; on balance Trypho's report was encouraging.

mentionem...factam I know of nothing to support the usual explanation: 'formula forensis quae usurpatur a nomine apud praetorem deferendo'. *Quinct.* 37 certainly does not: *doce te petisse ab eo...pecuniam, doce aliquando mentionem fecisse, dixisse deberi.* How then to reconcile the first statement with the second? I previously conjectured that something had dropped out, e.g. *mentionem* ⟨*P.*⟩ *nullam*, 'no mention of Publius (Clodius) as a possible prosecutor'. But *mentio* is used particularly of formal or public mention (cf. *Thes.* VIII. 775. 5), and very occasionally seems to imply such without further precision: cf. *Sest.* 85 *gladiatores ex praetoris comitatu comprensi...emissi a Serrano: mentio nulla*; *Verr.* II. 2. 46 *cuius hereditatis...controversia fuerat nulla, mentionem fecerat nemo*, 55 *ante quam aditum in ius esset, ante quam mentio denique controversiae facta esset ulla.* So here perhaps *mentionem* implies a definite, public declaration (not necessarily actual *postulatio*) as distinct from floating gossip.

5 **C. Clodi filio** C. Claudius Ap. f. Pulcher, Pr. 56, was the second of the three Pulchri: cf. 90 (IV. 15). 2 n. The spelling *Clodi* (cf. 90 (IV. 15). 2) may be contemptuous, if this was indeed a plebeian form adopted by P. Clodius for demagogic reasons (cf. W. M. Lindsay, *Latin Language*, p. 41). Gaius had two sons, both Appii; the younger, originally perhaps a Gaius, was adopted by his uncle Appius (Cos. 54) and so became known as Appius minor in distinction from the elder, Appius maior. This elder son may well be the Appius of 139 (VII. 15). 3. Later he was on good terms with C., who recommended him to D. Brutus in 43 (cf. *Fam.* XI. 22). However, he declared for the Triumvirs and received the Consulship in 38, followed six years later by a Triumph.

a fratre It is strange that C. gives no indication as to Quintus' views about the threatened prosecution. Presumably he wrote nothing on the subject, or nothing significant, apart from a tribute to Atticus (§3). Of course something like *laetiores* may have fallen out before *litteras*, or *isque* may represent *similisque* (proposed but not advocated by W. S. Watt, *Eranos*, 60 (1962), p. 71). *is...isque* is not very elegant, less so than e.g. 199 (X. 8). 10 *eo velim tam facili uti possim et tam bono in me quam Curione. is ad Misenum VI Non. venturus dicebatur.*

6 **Sesti** P. Sestius L. f., whose father had been Tribune early in the century. His career up to 56 is well known from C.'s defence *de vi*. As Tribune in 57 he took a leading part with his colleague Milo in opposing Clodius and promoting C.'s recall. Later Praetor and in 49 C.'s successor in

Cilicia, he went over to Caesar after Pharsalia, supported C. in 43 (*ad Brut.* 5. 4), but lived to sign senatorial decrees in 39 and 35. His second wife was a Cornelia of the Scipios (cf. 12 (I. 12). 1 n.). C.'s references to him in confidential letters tend, gratitude notwithstanding, to a certain irony, with which the notoriously wearisome quality of Sestius' literary style no doubt had something to do (cf. 141 (VII. 17). 2 n.).

9 **Appi** Cf. 42 (II. 22). 2 n. As Praetor-Designate he had been assigned the presidency of the *quaestio de repetundis*.

3, 2 **quod...quod** Cf. 21 (II. 1). 5 n. (*illam consularem*).

3 **nobis** I.e. 'to myself (restored) and my brother (freed from danger)'. *incolumes* = *salvi*, 'in full possession of civic rights'.

5 **vereare** Unlike Sallust and Catullus C. does not use the infinitive after *caveo*.

63 (III. 18)

The date is indicated by those of the preceding and following letters, 4 and 15 September.

1, 2 **Varronem** Cf. 60 (III. 15). 3.

pro amicitia Cf. 199 B (X. 8 B). 2 (Caesar) *pro iure nostrae amicitiae a te peto*; Sall. *Iug.* 9. 2 *tibi quidem pro nostra amicitia gratulor*; Prop. I. 22. 2 *quaeris pro nostra semper amicitia*.

4 **actorem** Someone to propose the necessary legislation: cf. 19 (I. 19). 4 n.

5–6 **aliquid in spe** Cf. 13 (I. 13). 6 n.

2, 1–2 **fac...me fac** The repetition may be pathetic. There is no need to change the first *fac* to *nunc* or *tu*. Baiter conjectured *fac me*, again needlessly: cf. with Sjögren 370 (XIV. 16). 3 *me fac ut expedias*.

3 **homo mirus** Take closely with *qui...amat*: cf. 300 (XIII. 29). 3 *o te ferreum qui illius periculis non moveris*. The idea that *mirus* is very rarely used of men can now be refuted from *Thes.* VIII. 1072. 45. Here it means 'wonderful', as (ironically) in 408 (XV. 29). 2 *mirus civis*, rather than 'extraordinary', as in 238 (XII. 2). 2 *o miros homines!*—not, of course, a wide distinction.

64 (III. 19)

1, 7 **natura** For the nominative see K.–S. I, p. 462. Lucr. IV. 984 *usque adeo magni refert studium atque voluptas* seems to be the only contemporary parallel. Normality can easily be restored by adding *quae esset* (Müller); but perhaps it was the inelegance of *interesset quae esset* that C. preferred to avoid.

lucem Cf. 52 (III. 7). 1 *lucem aspicere vix possum*.

8 **ex tuo portu** 'Non accipio μεταφορικῶς dictum, sed re vera ex portu...*ex tuo* autem, ex quo tu solvis cum in Italiam a tuis possessionibus redis' (Manutius).

9–10 **sustentabo...abiecero** Cf. Sjögren, *Eranos*, 16 (1916), p. 47: 'apparet futura...aequabiliter posita esse servata tamen utriusque stirpis proprietate, ut illud magis, quod durat, significat (*sic*) neque initii agendi neque finis ratione habita, hoc verbi notionem quasi in unum comprehendat atque complectatur': also K.–S. I, p. 148.

2, 1 **numquam sqq.** Cf. 55 (III. 10). 1 n.

2 **ac tamen** Cf. Madvig on *Fin.* II. 85.

6 **promissa** Made perhaps to Atticus: cf. 67 (III. 22). 2 n. *Lentulus suo in nos officio, quod et re et promissis et litteris declarat* sqq.

8 **et fidelis** To be taken closely with *obsecrationem.*

3, 1 **T. Pomponi** Cf. 74 (IV. 2). 5 *mi T. Pomponi* and 60 (III. 15). 7 n.

65 (III. 20)

POMPONIANO ATTICO Q. Caecilius (cf. 10 (I. 1). 3 n.) had just died. Atticus, being adopted in his will, took his name, adding as here appears both his former *nomen gentile* with the *-anus* suffix after the old style and his former cognomen after the new: cf. Varro, *R.R.* II. 2. 2 *Atticus, qui tunc T. Pomponius, nunc Q. Caecilius cognomine eodem* (cf., however, 67 (III. 22). 3 n.). M. Pupius Piso Frugi Calpurnianus would be almost analogous if he is correctly so called, but this is improbable: cf. 13 (I. 13). 2 n. It must be remembered that the cognomen in Atticus' case was personal and flatteringly descriptive.—According to Valerius Maximus (VII. 8. 5) public opinion was so outraged by Caecilius' conduct in leaving his benefactor L. Lucullus out of his will that his corpse was dragged through the streets.

1, 8 **diemque natalem** A favourite motif later: cf. 73 (IV. 1). 8 *alterius vitae quoddam initium ordimur*; 121 (VI. 6). 4 *propter hanc* παλιγγενεσίαν *nostram*; *post Red.* 27 *illo die quem P. Lentulus mihi...natalem constituit* (cf. *Sest.* 131); *post Red. ad Quir.* 5 *a parentibus...parvus sum procreatus; a vobis natus sum consularis.* So Sallust (*Orat. Cottae* 3) makes another returned exile describe himself as *bis genitus* (cf. Plaut. *Capt.* 891 *iterum natus videor*; *Bell. Afr.* 90. 3 *se eo demum die natos praedicantes*; and C.'s own *o fortunatam natam me consule Romam*). Recent notices from Rome had evidently lifted the despondency of 64 (III. 19).

10 **maxime** = *potissimum*, with *apud te in Epiro*: cf. 51 (III. 6) *nobis iter est in Asiam, maxime Cyzicum. tamen* seems to be elliptical, implying 'although I have not actually done so'. Atticus may be supposed to have referred to the subject in his letter. Since he wrote in reply to 60 (III. 15)—cf. §2 *de domo et Curionis oratione*—he may well have put in something like 'from your letter I hope to hear of you in Epirus soon' (cf. 60 (III. 15). 6). There is little to be said for *maxima, tamen* (Corradus).

2, 1 **de domo** Cf. 60 (III. 15). 6. C. writes to Terentia on the same day

(*Fam.* XIV. 2. 3) *quod de domo scribis, hoc est de area, ego vero tum denique mihi videbor restitutus si illa nobis erit restituta.*

1 **oratione** Cf. 60 (III. 15). 3.

5–6 **mihi gratum est** Almost 'I am grateful'. Atticus had doubtless written or implied that he had got his hands free the better to devote himself to C.'s affairs. The new paragraph which modern editors begin with this sentence obscures a logical nexus.

8 **multas partis** Cf. 92 (IV. 18). 2 *miram securitatem videbis, cuius plurimae mehercule partes sunt in reditu tuo*; *Verr.* II. 2. 10 *in hac causa multo pluris partis mihi defensionis quam accusationis suscepisse videor*; Lucr. III. 124 *noscere ut hinc possis non aequas omnia partis / corpora habere.* The metaphor is perhaps financial, from shares in a property or business venture.

3, 1 **quod me vetas** Cf. 60 (III. 15). 4.

7 **rogatio Sesti** A draft law, of which Atticus approved (cf. 68 (III. 23). 4), to repeal that of Clodius.

8 **et id** *et* del. Wesenberg; but cf. Lehmann, *Quaest.* p. 64.

9 **III Non.** *IIII* in editions before Constans is corrected in the light of his report: cf. his note, vol. II, p. 23. This letter will have been dispatched along with *Fam.* XIV. 2 to C.'s family, likewise dated 5 October. It is noteworthy that he does not say anything to Atticus about a painful, though to us obscure, incident involving Terentia of which he had heard from P. Valerius (*Fam.* XIV. 2. 2). *Kal.* (Sternkopf, followed by Sjögren) for *Non.* would make C. reply to Atticus' letter, received on 26 September (cf. 66 (III. 21) n. *triginta dies*), within two days; but he may have had to wait for a messenger.

66 (III. 21)

1 **triginta dies...ipsi** This expression probably means 'thirty whole days' (not including the day of writing or the day on which Atticus' last letter arrived): see L. W. Hunter, *Journ. Rom. Stud.* 3 (1913), pp. 94 ff.

3 **ibi...potissimum** For the order cf. 73 (IV. 1). 4 *ea...quae te puto potissimum ex meis litteris velle cognoscere*; *Fam.* III. 9. 3 *quonam te remunerarer potissimum genere*; Ter. *Andr.* 453 '*quem*' *inquit* '*vocabo ad cenam meorum aequalium / potissimum nunc?*'

5–6 **ut scribis** In reply to 60 (III. 15). 8.

67 (III. 22)

The first three paragraphs were written at Thessalonica, clearly at the same time as *Fam.* XIV. 1 to C.'s family. In mid-November he left for Dyrrachium, taking both letters with him, and added postscripts on arrival there.

1, 1 **Piso** Cf. 8 (I. 3). 3 n.

2 **occupatio** Perhaps with a hint of reproach: cf. 65 (III. 20). 2 *quod te ab omni occupatione expedisti, valde mihi gratum est.* In fact Atticus does not seem to have written again till 25 October (cf. 68 (III. 23). 1), five to six weeks after the letter answered in 65 (III. 20).

5–6 **non eadem** *eadem quae mihi* is negatived as equivalent to *cuius particeps sum.*

7 **milites** So to his family (*Fam.* XIV. 1. 3) *ego volebam loco magis deserto esse in Epiro quo neque Piso* (*ipso* codd.) *veniret nec milites, sed adhuc Plancius me retinet; sperat posse fieri ut mecum in Italiam decedat.* L. Piso left for Macedonia before the end of his consular year, *paludatus* (cf. *Pis.* 31; *Sest.* 71).

2, 1 **Lentulus** The Consul-Designate P. Cornelius P. f. Lentulus Spinther (C. never uses the second cognomen, said to have been derived from Lentulus' resemblance to an actor, but other contemporaries do) was probably like other Lentuli a patrician (on Willems' theory to the contrary cf. Münzer, *RE.* IV. 1394. 8) and brother of Cn. Lentulus Marcellinus, who succeeded him as Consul. As Curule Aedile in 63 he supported C., became Praetor in 60, and governor of Hither Spain the following year. At this time he was on good terms with Caesar (cf. Caes. *B.C.* I. 22 f.) as well as Pompey. As Consul his leading part in promoting C.'s recall earned him such effusions of gratitude as *post Red.* 8 *parens ac deus nostrae vitae, fortunae, memoriae, nominis.* His later fortunes included a proconsulate in Cilicia (56–54), a Triumph (51), capture and release by Caesar at Corfinium, Pharsalia (either he or Lentulus Crus commanded the Pompeian right), and death in Africa, perhaps by Caesar's orders. Nine letters addressed to him almost fill the first book *ad Familiares.* They are in C.'s most orotund style, abounding in compliment and without any flavour of intimacy. Nothing in his record suggests that his industrious mediocrity as an orator (*Brut.* 268) did not extend to his whole personality.

promissis et litteris The promises presumably being to C.'s friends in Rome (cf. 64 (III. 19). 2 n.), the letters to C. himself. In *post Red.* 20 *cuius* (sc. *P. Sesti*) *ego clientibus, libertis, familia, copiis, litteris ita sum sustentatus ut meae calamitatis non adiutor solum verum etiam socius videatur* I suspect that *litteris* refers to published manifestoes ('compositions') rather than to private letters.

3 **eum** Lentulus, of course.

4 **Metello** Cf. 57 (III. 12). 1 n.

quantum speraret 'How much he hoped' rather than 'as much as he hoped', which would seem to require *sperasset* (Wesenberg).

3, 1 **mi Pomponi** Here as in 68 (III. 23). 2 and 74 (IV. 2). 5 the adoption is ignored. In later letters to Atticus C. drops the *nomen gentile* in favour of the cognomen, except in 130 (VII. 7). 7 *adsentior Cn. Pompeio, id est T. Pomponio*;

but in other correspondence and in dialogues he uses both. He never calls Atticus by his new nomen Caecilius or his new praenomen Quintus, but calls his daughter Caecilia in 118 (VI. 4). 3.

3 **cum desiderio** This makes a much more natural way of interpreting the phenomena in the MSS than the repellent asyndeton bimembre *luctu, desiderio* in Sjögren's and other texts, followed by *cum omnium rerum ⟨tum meorum⟩ qui* (Baiter). Cf. 64 (III. 19). 3 *si me omnibus amplissimis, carissimis iucundissimisque rebus...spoliatum...vides.*

4, 2 **eram auditurus** Since the postscript to *Fam.* XIV. 1 was certainly added at Dyrrachium, so no doubt was this. Otherwise the tenses might be epistolary, with everything but the date written before departure from Thessalonica.

2 **et quod** The following *et* is probably a mere dittography. Sjögren suspects that *et officiosos* or the like has fallen out, comparing 337 (XIII. 45). 3 *nec diligentiorem nec officiosiorem nec nostri studiosiorem* and *Fam.* XIV. 1. 7 *Dyrrachium veni quod et libera civitas est et in me officiosa et proxima Italiae.* But *officiosos* here would add nothing of value to *mei studiosos.*

5 **cuicuimodi** Cf. Neue–Wagener, *Formenlehre*[3], II, p. 513.

6 **aut rem...exspecto** An epigrammatic way of saying, 'I expect results now, and if I don't get them I don't expect (i.e. don't want) any more hopeful talk'. Cf. *Dom.* 142 *aut re aut spe*; *Orat.* 107 *non tam re...quam spe* et sim.

68 (III. 23)

1, 6 **octo** The names of four are known: L. Antistius, L. Ninnius Quadratus, L. Novius, Q. Terentius Culleo (cf. Broughton, pp. 195 ff.). Clodius and Aelius Ligus were the two dissidents (cf. *Sest.* 69). One of them no doubt vetoed the law, but Broughton's statement that Ligus did so is not attested definitely (he had vetoed a senatorial decree proposed by Ninnius on 1 June: cf. *Sest.* 68).

9–10 **si iam...desperata erit** Promulgated on 29 October, the law could have come before the Assembly on 21 November at the earliest: cf. Mommsen, *St.* III, p. 376, n. 1 ('gegen Cicero' is a slip). News of its failure could hardly have reached Dyrrachium by the 29th, but C. (and Atticus) could expect that it would be vetoed; his words do not imply that this had happened to his certain knowledge. It is unlikely that a veto was exercised before the voting day—that would have been unusual if not actually illegal (*ibid.* I, pp. 284 f.)—though it may well have been threatened. Clodius in his speech on 3 November (cf. §4) seems to have taken for granted that the bill could not pass.

13 **nostris** Cf. §3 *octo nostri tribuni pl.*; 111 (V. 18). 3 *amicos consules habemus, nostrum tribunum pl. Furnium.* The text can stand, though *novis* (Corradus) is tempting in view of *veterum* in the next sentence. Winstedt

and Constans, both reading *nostris*, translate 'our new magistrates', 'les nouveaux magistrats'.

2, 3 **civitatem** Despite C.'s argumentation in *Caec.* 95 ff., *interdictio aqua et igni* was generally held to deprive a Roman of citizen status.

ordinem I.e. membership of the Senate.

4 **quae cavenda fuerint** In respect of house and property: cf. 60 (III. 15). 6.

8 **scis** Apparently an admission on C.'s part. *enim* must therefore be taken as starting the whole exposition (cf. 16 (I. 16). 1 n.). Or did C. write *scio*?

13–14 **quo non...oporteat** So Malaespina conjectured and so Bosius read, 'auctoribus libris manuscriptis; in quibus manifeste scriptum erat *abrogatur* et *quo non* pro *abrogaretur* et *quo modo*, quae lectio est excusorum'. For the position of the negative cf. 117 (VI. 3). 4 *equidem doleo non me tuis litteris certiorem fieri.* But *quo minus* (Klotz) or *quo...non oporteat* would certainly be more normal Latin. The vulgate *quo modo* is nonsense.

3, 5 **abrogare sqq.** Cf. *Rep.* III. 83 *huic legi nec obrogari fas est neque derogari ex hac aliquid licet neque tota abrogari potest.* Ancient authorities differ slightly in their definitions of *obrogare*, but the root meaning is clear: 'enact in contravention of a prior enactment (without actually repealing or amending it)'.

5–6 **non liceat, non licuerit** The tenses apparently refer to the times of *rogatio* and *promulgatio* respectively, so as to include legislation passed between the two.

7 **E.H.L.N.R.** *eius hac lege nihil rogatur.*

4, 1–2 **lege...non tenebantur** Evidently a statement of general application. Mommsen (*St.* I, p. 291, n. 3) finds support for it in Nero's refusal as Consul in A.D. 55 to allow his colleague to take the oath *in sua acta* (Tac. *Ann.* XIII. 11. 1).

8–9 **quod si opus esset** *quo* Madvig; but *quod* is a conjunction.

9 **Ninnium** L. Ninnius Quadratus, perhaps of noble Campanian family, is warmly eulogized in *Sest.* 26 and elsewhere for his opposition to Clodius. In 55 he is found opposing the lex Trebonia along with Cato and Favonius (Dio, XXXIX. 35. 5), and in 49 seducing C. into armed resistance to Caesar (200 (X. 16). 4).

11 **sibi** Emphatic. For the inversion cf. *Cael.* 18 *reprehendisti a patre quod semigrarit*; J. Marouzeau, *L'ordre des mots dans la phrase latine*, III, pp. 121 ff. Palaeographically it is as easy to suppose with Lehmann that a clause beginning *sive* (or rather *sive quod*) has fallen out, but it is not so easy to invent its content. *scilicet* (Lallemand) or Hofmann's *sane* (= *nimirum*) for *sive* make appropriate sense, though I am not quite happy about the Latinity of the latter despite Ter. *Eun.* 89, 785.

12 **id caput** In Clodius' law, forbidding a relevant motion or discussion in the Senate (cf. 60 (III. 15). 6). Others besides the Tribunes seem to have ignored it (*ibid.*).

20 **T. Fadio** T. Fadius (on his supposed cognomen Gallus see *Cl. Rev.* N.S. 12 (1962), pp. 195 f.) was C.'s Quaestor in 63 and Tribune in 57. After 52 he was in exile for reasons unknown and the probable recipient of one of C.'s most disagreeable letters (*Fam.* VII. 27), as well as of the conventionally consolatory *Fam.* V. 18.

Visellius C. Visellius C. f. Varro (Corradus conjectured the right praenomen, but *T.* is probably mere repetition) was the son, perhaps adoptive, of the jurist C. Visellius Aculeo and C.'s aunt Helvia. Quaestor before 73, Curule Aedile *c.* 59, and (according to his cousin) a talented though not popular speaker (*Brut.* 264).

Sesti Cf. 65 (III. 20). 3.

5, 4 **studiis bonorum** W. S. Watt (*Eranos*, 60 (1962), p. 72) would read *studiis* ⟨*tribuniciis*⟩, *bonorum* sqq., comparing 69 (III. 24). 1 *studia tribunorum*, an attractive suggestion; but the vulgate is intelligible, if not very elegant.

multitudine comparata Cf. *Sest.* 127 *mihi negas optandum reditum fuisse per familias comparatas et homines armatos.* Clodius' taunt seems to have had some substance: cf. 74 (IV. 2). 7 n.

8 **miserum misere** Such expressions are common in Plautus: e.g. *Aul.* 721 *heu me miserum, misere perii*, *Amph.* 278 *optumo optume optumam operam das.* Cf. Munro on Lucr. III. 898 *misero misere*; J. B. Hofmann, *Lat. Umgangssprache*², § 88; H. Haffter, *Untersuchungen zur altlat. Dichtersprache*, ch. I.

9 **gravius consulere** Evidently an allusion to the possibility of Quintus' committing suicide. C.'s intention to do so is implied throughout this appeal, especially in what he says about his son.

12 **Terentiam** The omission of Tullia from the list is odd.

13 **primorum** Following the entry of the new Tribunes into office on 10 December, say commentators; so to his family (*Fam.* XIV. 3. 3) *nunc spes reliqua est in novis tribunis pl. et in primis quidem diebus.* But § 1 *hortaris ut... mensem Ianuarium exspectem* suggests that he had the new year in mind: cf. *post Red.* 5.

14–15 **se...dederint** Cf. Enn. *ap. Nat. Deor.* III. 66 *ita dat se res ut operam dabit*; Ter. *Hec.* 380 *ut res dant sese.*

69 (III. 24)

1, 1 **vestro** Including, doubtless, Atticus, Quintus, and C. Piso Frugi. There appears to have been an organized group working for C.'s recall to which some well-wishers outside his family, such as Sestius, probably belonged.

2 **ornatas** The consular provinces, Cilicia and Hither Spain in the present case, were assigned by the Senate under the lex Sempronia in advance of the election, but the decree for their military and civil provision (*ornatio*: cf. Mommsen, *St.* III, pp. 1097 ff.) was usually passed after the Consuls had entered upon office, and could be held up by tribunician veto. Why special treatment, unprecedented as C. believed (§2), should have been accorded to the Consuls of 57 we do not know.

3 **aliquid aliquando** A favourite collocation: cf. e.g. *Verr.* I. 5 *fortasse aliqua in re nos aliquando fefellisset*, II. 1. 44 *quorum simile forsitan alius quoque aliquid aliquando fecerit*; *Brut.* 209 *cum autem difficile sit in longa oratione non aliquando aliquid ita dicere* sqq.

6 **videtur** *videretur* Lambinus, and editors prior to Sjögren. *videbatur* would seem necessary if a historic tense were required, which it clearly is not.

9 **omnem vim...amiserint** Since the *ornatio* had been put through before their entry into office on 10 December the new Tribunes had not been able, as they normally would have been, to influence the proceedings by exercising or threatening their veto.

14 **se invitis** It is not to be assumed that any of the Tribunes-Designate in writing to C. or talking to other correspondents actually threatened to turn against him, though two of them, Numerius Rufus and Atilius Serranus, in fact did so; but doubts may have been expressed about the reactions of the College generally.

16 **nullo modo potuit** Presumably C. means that there would not have been time to pass a law in the Assembly before 10 December, after which it would be subject to veto by the new Tribunes.

2, 2 decernere Cf. *post Red.* 6 *postea* (i.e. after the failure on 1 January) *nihil vos civibus, nihil sociis, nihil regibus respondistis...nihil hic ordo auctoritate declaravit.* The infinitive corresponds to a present indicative (or epistolary imperfect), future in sense, *non decernitur*: similarly 81 (IV. 12) *vix videor praesto esse*; 132 (VII. 9). 3 *idque eum...non facere miror*; 181 A (IX. 13 A). 1 (Caesar) *cum in spem venero de compositione aliquid me conficere*; 300 (XIII. 29). 1 *ut mihi ille emptor non esse videatur*; *Fam.* I. 7. 10 *quae ne per populum quidem sine seditione se adsequi arbitrabantur*, II. 11. 1 *quod videmur effugere si ad constitutam diem decedemus*; Plaut. *Asin.* 442 *aibat reddere quom extemplo redditum esset*; Caes. *B.G.* II. 32. 3 *illi se quae imperarentur facere dixerunt. posse* should not be added in such cases.

6–7 **quae...suscepta** '*qua* Orelli, vulg. temere. *constantiam suscipere* non minus recte dicitur quam *animum, mentem voluntatemque, dolorem, benevolentiam, cogitationem*' (Müller). Perhaps this is not quite true since *constantia* denotes action (perseverance in a course of behaviour) rather than a mental attitude or intention. But the extension, if it can be called one, is surely not too difficult.

7–8 **iam iis** *iam* introduces a new point: cf. *Thes.* VII. 121. 22. But here as in 115 (VI. 1). 13 *iam Scrofa vellem haberet ubi posset* elliptical *nam* would be so appropriate that corruption may be suspected.

8 **relatum est** In the senatorial debate.

8–9 **erat...difficile** The indicative is regular: cf. K.–S. I, pp. 171 ff.

10 **omnino** Concessive (not 'de toute façon').

12–13 **tamen...potuerimus** But could they? Atticus on the spot was likely to be a better judge than C. *tamen* implies 'even if we had not agreed'.

13 **amiserimus** Cf. §1 *ne...studia tribunorum amiserimus*. Defence of *miserimus* (cf. Sjögren, *Comm. Tull.* p. 148) is idle. *Fin.* I. 2 *difficilem quandam temperantiam. postulant in eo quod semel iam missum coerceri reprimique non potest* is an equestrian metaphor (cf. Prop. III. 11. 62 *at Decius misso proelia rupit equo*).

14 **quo loco** Usually in such cases C. writes *quo loci*, as 140 (VII. 16). 3 *quo loci sit res*; 13 (I. 13). 5 *res eodem est loci quo reliquisti*; but the MSS have *loco* in *Verr.* II. 5. 164 *quo loco haec causa sit*.

14–15 **ut instituisti** 'As you have (recently) begun to do.'

70 (III. 25)

1 **datae** The case against *a me* is pretty well conclusive. (*a*) If, as Sternkopf argued (*Philol.* 66 (1907), pp. 315 ff.), Atticus met C. at Dyrrachium on his way to Epirus, they would surely have remained together; the hypothesis that Atticus did not wish to witness C.'s despair, unconvincing anyway, could not apply if the letter or letters which caused the despair arrived after they had met and parted. At any rate we should expect C. to join him presently, instead of making him promise to come back to Dyrrachium, or wherever else C. himself might be, before 1 January. (*b*) As W. S. Watt emphasizes (*Eranos*, 60 (1962), p. 73), *discessisses* below must refer to Atticus' departure from Rome; and it is hard, if not 'impossible', to take *discessum* and *discessisses* with reference to different journeys. *datae* is a great improvement on earlier suggestions (*ab urbe, a meis*); clearly the time of dispatch (if *discessum* means departure from Rome) is what matters, not the time of arrival. Watt explains the corruption by supposing that the first letter was lost and *a te* 'emended' to *a me* on grounds of sense. Still a difficulty remains. If Atticus had set out for Buthrotum C. was presumably writing to him there: but why did C. not go to Buthrotum himself? From the following letters they seem to have stayed apart at any rate until February, nor do these suggest that Atticus was in Greece. Perhaps then he had left Rome on some other errand (cf. 72 (III. 27) n.).

1 **litterae** From Quintus and/or others in Rome.

2 **perspicio** There seems to be no evidence for *prospicio* with accusative and infinitive.

calamitate Cf. 58 (III. 14). 2 n.

5–6 **ne...videamur** C. does not wish to appear as willing that Atticus' affairs should be neglected because his own were hopeless.

7 **te...sistas** Cf. 208 (X. 16). 6 *te vegetum nobis in Graecia siste*. In legal vocabulary *sistere* (intransitive), *sisti* or *se sistere* = 'appear in court'.

71 (III. 26)

There is nothing to show that this and the following note were not written from Dyrrachium, though it is just possible that C. had already moved to Buthrotum (cf. 72 (III. 27) n.). Quintus' letter will have arrived towards the middle of January, and C. obviously wrote soon afterwards.

1 **litterae** On 1 January the Senate was proceeding to vote a decree in favour of C.'s recall (cf. *Sest.* 74 *cum...fieretque sine ulla varietate discessio*—note the tense), when a now hostile Tribune, Sex. Atilius Serranus Gavianus (cf. 74 (IV. 2). 4 n.), got up and asked for an adjournment, so that he could decide overnight whether to use his veto or not. This being conceded, he managed at the next and subsequent meetings to put up further obstructions (*ibid.*). Quintus' letter appears to have reported only the proceedings on 1 January and to have reported these inaccurately, unless C. misunderstood; for *factum* implies that the decree was actually passed (cf. Caelius, *Fam.* VIII. 8. 4 *senatus consultum, quod tibi misi, factum est auctoritatesque perscriptae*), whereas the account in the *pro Sestio* shows that Serranus succeeded in holding it up, without having to exercise his veto.

2 **legum** Cf. *Sest.* 74 *Cn. Pompeius...dixit sese...censere ut ad senatus auctoritatem populi quoque Romani beneficium erga me adiungeretur*. One of the laws in question was presumably the tribunician *rogatio* of Fabricius (see on 71 (III. 26)), promulgated not later than 29 December.

3 **auctoritate** Cf. *Sest.* 74 above and 75 *cum omni mora, ludificatione, calumnia senatus auctoritas impediretur*, unless *auctoritas* has there its general sense (cf. 16 (I. 16). 1 n.). Here *utar* seems to show that C. refers to the actual resolution. *auctoritas* in this use generally denotes a vetoed decree; this one was not vetoed, and though its passage was thwarted C. did not know that. Occasionally however the term is used 'bei der eminenten Manifestation der politischen Machtstellung des Senats': cf. Mommsen, *St.* III, p. 1033, n. 2.

72 (III. 27)

On 23 January 57 the Tribune Q. Fabricius brought a law for C.'s recall before the Assembly, but its passage was thwarted by Clodian violence.

Q. Cicero narrowly escaped death in the ensuing riot (cf. *Sest.* 75 ff.; Dio, XXXIX. 7. 2). A letter from Atticus, who seems to have returned to Rome in January (cf. 70 (III. 25). 1 n. (*datae*)), would reach C. early in February.

1 **ex re ipsa** *et* could be dispensed with: cf. *Font.* 25 *non solum de reo, sed etiam de accusatore, de teste iudicare.*

3 **cito videbo** Atticus had not kept his promise to join C. before the beginning of the year. The cessation of their correspondence until C.'s return to Rome in September suggests that he did so in February, and that they spent the ensuing five months or so together at Buthrotum. This view of Constans seems more plausible *per se* than Bardt's surmise (*Röm. Charakterköpfe in Briefen*, p. 173, n. 1) that Atticus suppressed the remaining letters from exile because they revealed a *Verstimmung* between the two friends. The fact that C. took ship from Dyrrachium (73 (IV. 1). 4) does not, of course, prove that he had been there all the time. On the other hand C.'s language in 73 (IV. 1). 2 reads rather oddly if he and Atticus had been together from February to July. I do not see any way to settle the question.

73 (IV. 1)

With Caesar's consent (cf. *Fam.* I. 9. 9; *Pis.* 80) and Pompey's active support, the movement for C.'s restoration finally prevailed. In July, pursuant to a decree of the Senate, the whole body of magistrates except the Praetor Ap. Pulcher and the two hostile Tribunes (*Pis.* 35) promulgated a bill, which was unanimously passed by the *comitia centuriata* on 4 August. That same day, leaving Atticus in Greece, C. sailed for Brundisium, where he stayed with his host of the previous year, M. Laenius Flaccus (*Sest.* 131). What followed let himself relate (§§ 3 ff.). His letter seems to have been written shortly after the meeting of the Senate on 8 September (§ 7).

1, 3 **gratularer** 'Congratulate'; Atticus would rejoice in the success of his own efforts. The meaning 'thank' is not established in C.

cognoram Applies in the first clause (*te... diligentem*) to events immediately prior to C.'s exile, in the second (*eundemque... contulisse*) to the exile itself. The passage needs to be read in conjunction with 60 (III. 15). 4 and 7.

5–6 **nec etiam... diligentem** Lit. 'and not, furthermore, over-diligent by reason of attentions on my part', etc.: C. implies that Atticus *was* insufficiently sedulous, but only because he, C., had not done enough in the past to make him more so: *nimium* practically = *valde*, and *propter... observantiam* belongs closely with *diligentem* (sc. *factum*). This is not very neat writing; but even C.'s powers of expression were not quite equal to the strain here placed upon them. How, at this solemn moment, was he to render Atticus his full due of gratitude and affection while making it plain that earlier inadequacy, as C. saw it, had not been forgotten? The effort to say an essentially ungracious thing tactfully seldom conduces to smoothness and lucidity of style. For the sense cf. 60 (III. 15). 4 *quod meritis meis perfectum potuit, ut dies et noctes quid mihi faciendum esset cogitares, id abs te meo non tuo scelere praetermissum est.* So only I think is the text explicable; certainly not by taking *observantiam* with Sternkopf and others as 'lack of observance'. Two conjectures are worth mention: (*a*) *pro praeterita mea in te observantia* (Bosius). This is palaeographically presentable, though Bosius' report of Z must be discounted (cf. Intr. p. 94). But the stricture on Atticus becomes unduly harsh, and flatly contradicts admissions in 60 (III. 15). 4 and 65 (III. 20). 3. The ambiguity of *praeterita*, which might mean 'neglected' (cf. *Quinct.* 68 *officium...praeteritum*) is a further objection; indeed *praeterita* (without *pro*) would make a possible reading in this sense. (*b*) *propter meam in te* ⟨*in*⟩*observantiam* (Madvig). That the noun does not occur in extant Latin before Quintilian is no objection; many such negative forms are used by C. only once or twice. But in so delicate a context the phrase lacks finesse.

6 **eundemque** Strongly adversative, as T.–P. say.

2, 4 **quem** I.e. *et te* (understood from *tuum*).

si umquam Not only a stylistic improvement on *numquam* but in line with the indications of the MSS. *s-* fell out after *-us*, leaving *iumquam*, hence *tum quam*.

5 **praeteriti temporis** Cf. *Sest.* 31 *meam causam praeteriti temporis*, and for the double genitive *Flacc.* 5 *longinqui temporis cognitarum rerum fidem*; *Leg. Agr.* III. 2 *parvam exigui temporis usuram bonae de me opinionis*; K.–S. I, p. 416.

3, 1 **statu** 'Position' in the community at large.

2 **forensem** 'Public' as opposed to 'private': cf. §7 *forensium rerum.* C. can hardly be referring to his reputation as an advocate in a letter written only a few days after his return.

4 **optaramus** If C. had wished to convey 'in greater measure than I desire' he would have written *volumus* or *vellemus*, not *optamus*; neither is it credible that he would have denied wanting influence with the *boni* at this stage, not to speak of senatorial and public prestige.

4, 4 **prid. Non. Sext.** Cf. throughout this passage *Sest.* 131.

6 **natalis** Brundisium was founded as a Latin colony in 244 (Vell. I. 14. 8) or a little earlier (cf. Liv. *Epit.* XIX).

7 **Salutis** Her temple on the Quirinal was dedicated in 302 by C. Junius Bubulcus (Liv. X. 1. 9). Atticus' house, the *domus Tamphiliana* recently inherited from his uncle (Nep. *Att.* 13. 2), was near by: cf. 290 (XII. 45). 2 n.

9 †VI Letters normally took from seven to nine days to reach Brundisium from Rome, a distance of 360 miles; 236 (XI. 21). 1 shows one arriving on the seventh day of travel. Five days seems incredibly short, particularly as the messenger could not have made an early start on the 4th.

Ten, on the other hand, is rather leisurely in the circumstances, so that *III* (Schmidt) is preferable to Constans' deletion of the numeral. I can see no significance as to date in *cum Brundisi essem*.

12 **honestissime** Sjögren compares *celerrimis* (due to *creberrimis*) for *celerrime* in *Fam.* III. 5. 4. Alternatively add *decretis* (Orelli).

5, 3 **se inimicos esse** Unwarrantably suspected as a gloss (Corradus).

5 **templorum** Including the great temple to Honos and Virtus: cf. Pease on *Nat. Deor.* II. 61.

infima plebe 'The lower orders', little worse than *vulgo*: Sjögren compares 370 (XIV. 16). 2 *infimorum* and 371 A (XIV. 17 A) *cum bonis omnibus tum infimo cuique*. Cf. also Liv. XXXIII. 46. 6 *infimorum quoque libertati gravem esse superbiam eorum*; Ascon. 45. 17 *quo facto et plebem infimam offenderat et senatus magnam gratiam inierat.* No need then for Lehmann's *ab infimo*, even if the sense were satisfactory, which seems to me dubious. *infima plebs* is found elsewhere (Ascon. 32. 19; Curt. X. 7. 1; Tac. *Hist.* II. 38; Suet. *Otho*, 7. 1), also *infima multitudo* (*Mil.* 95, etc.).

9 **Non. Sept.** I.e. *Nonarum Septembrium*.

gratias egimus In the speech *post Reditum in Senatu*.

6, 1 **eo biduo** On 7 September: cf. K.–S. I, pp. 356 f. Take with *feci et...dixi* at the end of the sentence.

caritas The price of corn had been high in early July (Ascon. 48. 20), but had dropped subsequently, only to soar again after C.'s return to Rome (*Dom.* 11, 15), probably as a result of the influx from the country.

1–2 **ad theatrum** The *ludi Romani* (5–19 September) were now in progress.

3 **mea opera** A retort to C.'s friends, who had attributed the price fall to divine approval of his recall and the sense of security it promoted (*Dom.* 15: cf. *post Red. ad Quir.* 18).

3–4 **per eos dies** On 6 September the Senate met in the Temple of Concord: cf. *Dom.* 11 *concursus est ad templum Concordiae factus, senatum illuc vocante Metello consule.* C., who was unwell (at least, he says he was unwell on the 7th (*ibid.* 15)), probably did not attend. The sitting of the 7th was on the Capitol. C. appeared in response to popular demand, but the Consulars opposed to Pompey made themselves scarce: cf. *Dom.* 8 *at enim non nulli propter timorem, quod se in senatu tuto non esse arbitrabantur, discesserunt.*

5 **bonorum** The generality of right-minded persons, not always at one with the *Catoniani*; so *Dom.* 16 *flagitabar bonorum expostulatione.*

6 **nominatim** Cf. *Dom.* 15 *a populo Romano universo, qui tum in Capitolium convenerat, cum illo die minus valerem, in senatum nominatim vocabar.*

7 **decernerem** Properly the Senate 'decreed' on the motion of the presiding magistrate, but the word is used of individual Senators originating or supporting a proposal, as in *Phil.* XIV. 29 *decerno igitur...supplicationes.*

7 **feci** I.e. *decrevi*. So often, e.g. *Tusc.* I. 24 *evolve diligenter...feci mehercule.*

8 **negarent** See K.–S. II, pp. 200 f.

9 **Messallam** Still a supporter of the 'Triumvirs': cf. 13 (I. 13). 2 n.

12 **cum multitudo** Slightly more probable graphically than the vulgate *cum populus* (ς). *continuo* may represent *c̄titudo*, *mul-* having fallen out. But it seems safer to retain it, assuming 'parablepsy' from *-tinuo* to *-titudo* (see, however, below).

13 **meo nomine recitando** Probably as a signatory to the decree. On the construction cf. K.–S. I, pp. 752 f. *in* was added before *nomine* by Zumpt, needlessly. But I rather suspect that the right reading may be *quo...recitato,* ⟨*cum multitudo*⟩ *continuo...plausum meo nomine iterando dedisset* sqq.

contionem Not the extant speech *post Reditum ad Quirites*, which says nothing about a corn shortage (quite the reverse, § 18) or Pompey's appointment.

14 **eam** More likely than *quam* (Wesenberg) to have fallen out after *-em*. Neither addition of course is absolutely necessary.

15 **dederunt** Cf. 14 (I. 14). 1 n. The three dissentients were doubtless once again Ap. Pulcher and the two Tribunes, Serranus and Numerius Rufus.

7, 1 senatus frequens A common phrase for a well-attended meeting, used, however, occasionally in a technical sense to denote a particular kind of meeting, perhaps one which required a quorum or in which penalties were imposed for absence: cf. 415 (XVI. 7). 1; *Phil.* III. 19; Liv. XXVIII. 9. 5; J. P. V. D. Balsdon, *Journ. Rom. Stud.* 47 (1957), pp. 19 f. The previous day's sitting had evidently been a thin one. That and the alleged difficulty of free expression will have formed grounds for a proposal to rescind the decree then passed (cf. *Dom.* 10; J. Stroux, *Philol.* 93 (1938), p. 87).

consulares Sc. *fuerunt*: cf. *Q.Fr.* II. 1. 1 *senatus fuit frequentior quam putabamus esse posse...consulares nos fuimus et duo consules designati, P. Servilius* sqq.

2 **quindecim** Both Ciceros were among those appointed. The others are unknown.

6 **Messius** C. Messius, probably of Campanian family, was Tribune this year. He had put forward an abortive bill for the recall of C. (*post Red.* 21), who successfully defended him in 54 (90 (IV. 15). 9), as Münzer suggests (*RE.* XIX. 1243. 31) on charges connected with his election to the Aedileship for 55: cf. 90 (IV. 15). 9 n. His friendship with Pompey (cf. 161 D (VIII. 11 D). 2) did not prevent him from serving as one of Caesar's Legates in 46 (*Bell. Afr.* 33. 2, 43. 1).

8 **nostra** The law was doubtless on the lines suggested by C. the previous day (cf. § 6). On its provisions cf. J. P. V. D. Balsdon, *Journ. Rom. Stud.* 47 (1957), pp. 16 ff.

10 **duce Favonio** Ironical. Favonius was at most a *tribunicius* in 57.

11 **pontifices** To them it fell to determine whether Clodius' consecration of part of the Palatine site to Liberty (cf. 60 (III. 15). 6 n.) was valid. C. addressed them in the extant speech *de Domo Sua* on the last day of the month. The matter had presumably been in their hands since 4 August.

responderant *responderunt* (ς) is usually read, but cf. 2 (I. 6). 2 *Quintus...quo volumus animo est in Pomponiam et cum ea nunc in Arpinatibus praediis erat.*

12 **habemus** *habebimus* is probably another *Falschverbesserung*; for the present cf. K.–S. I, pp. 119 f. The change to the future (*aestimabunt*) is natural, since the first consequence would be immediate, the second more remote.

13 **sin aliter sqq.** If the consecration is nullified C. will have the site on which to rebuild and receive compensation for his demolished house; that Clodius' structures would in this case be removed (cf. 74 (IV. 2). 5) is taken for granted. If on the other hand the consecration is upheld, he will be paid compensation for both site and building. The temple, as 'a conspicuous reminder of the Senate's defeat', will none the less be demolished and another, not associated with Clodius, built in its room. So R. G. M. Nisbet (*Cl. Rev.* N.S. II (1961), p. 240), whose defence of the text I accept against my former view, also that of Schütz and Boot, that something has fallen out (cf. *Towards a Text*, p. 15: the conjecture there advanced was anticipated by Schütz). Objects for *demolientur* and *locabunt* would make the meaning clearer (Clodius' portico (cf. 74 (IV. 2). 3 n.) as well as the temple would have to be replaced); but a reader *au fait* with all the circumstances might be expected to understand.

8, 1–2 ut...bonae The source of this line (Ribbeck, *Trag. Rom. Frag.*[3] Incert. 260), also quoted in 74 (IV. 2). 1 and *ad Brut.* 18. 2 (...*fluxa*...*bona*), is unknown. Shuckburgh rendered with Milton's 'For happy though but ill, for ill not worst'.

3 **domestica** Probably concerning Terentia: cf. 74 (IV. 2). 7 n.

6 **alterius vitae** Cf. 65 (III. 20). 1 n. (*diemque natalem*).

74 (IV. 2)

C. obviously wrote soon after the Senate's meeting on 2 October, reported in §5.

1, 2–3 non modo...sed ne...quidem Cf. K.–S. II, p. 62.

4 **iter** Cf. *Q.Fr.* III. 3. 4 *neque ego impediri Ciceronis iter atque illam disciplinam volo*; Plancus, *Fam.* X. 8. 2 *expeditius iter communis salutis quam meae laudis esse volui* (less similar).

9 **ut...bonae** Cf. 73 (IV. 1). 8.

2, 3 pontifices Cf. 73 (IV. 1). 7 n.

5 **dolor** *dolor* is sometimes hardly more than emotion (πάθος) as opposed to *ingenium*: cf. *Propertiana*, p. 21. The alternative reading *doloris magnitudo* (cf. *Thes.* v (i). 1853. 36) would to my feeling be less natural of emotion than of physical pain.

6 **nostrae** I.e. *Romanae*: cf. Prop. II. 3. 33 *hac ego nunc mirer si flagret nostra iuventus?* Or does C. mean 'devoted to me' (cf. 21 (II. 1). 3 *adulescentulorum studiis excitati*)?

deberi Cf. *Leg.* II. 7 *tibi horum nihil deberi potest*; *Tusc.* II. 67 *tibi hoc video non posse deberi*; *Top.* 4 *non potui igitur tibi...debere diutius* et sim.

3, 4 **mihi** Cf. §4 *senatui placere mihi domum restitui*. Such quotations do not always reproduce the original exactly: in 60 (III. 15). 6 *ne referri neve dici liceret* everyone recognizes that the wording is based on that of Clodius' bill, but it has been modified to suit what precedes. *emi* (Constans) is too clever, and an abbreviation *M.T.* (Ursinus) = *M. Tullio* (cf. *Tull.* 29; *Dom.* 44, etc.) lacks warrant.

6–7 **in contionem escendit** Commentators quote Gell. XVIII. 7. 5 (from Verrius Flaccus) *contionem autem tria significare, locum suggestumque unde verba fierent, sicut M. Tullius in oratione quae inscripta est contra contionem Q. Metelli 'escendi' inquit 'in contionem'* sqq. *escendere* is constantly changed to *ascendere* in MSS: cf. *Thes.* v (ii). 856. 63.

7 **nuntiat** A remarkable array of extravagant conjectures (see Müller *ad loc.*) suggests that *iam* is best deleted. It may have arisen from *-iāt*.

10 **Libertatem** The statue of the goddess of course stood in her temple (cf. *Dom.* 116 *ista Tanagraea oppressa libertate Libertas*), but Clodius need not be taken as referring to the statue specifically. His choice of this divinity had probably symbolized deliverance from C. the tyrant (*rex*): cf. W. Allen, *Trans. Am. Phil. Assoc.* 75 (1944), pp. 1 ff.

ut Perhaps to be omitted. *vi* (Corradus) is tempting, but the notion of *vim vi depellere* would probably have been expressed more epigrammatically if intended at all.

infimi *infirmi* should be valued on its own demerits. The MSS count for nothing here: cf. 370 (XIV. 16). 2; *Thes.* VII. 1446. 11.

12 **statueram** *statuerem* (Meutzner) would require emendation. The parenthesis explains *amentiam*.

13 **porticum Catuli** Erected by Q. Lutatius Catulus, Cos. 102, on the site of the house of C. Gracchus' associate M. Fulvius Flaccus. Clodius had destroyed it, at least in part, and replaced it by one which extended into C.'s ground (cf. 60 (III. 15). 6 n.).

4, 1 **adhibentur** The word seems carelessly chosen. As Senators these Pontifices would have the right and duty to attend anyway, and so could hardly in strictness be 'called in'. They spoke, as appears below (*loco*), in the normal order of official precedence. Possibly, however, they were specially summoned under penalty for non-attendance: cf. Mommsen, *St.* III, pp. 915 ff.

1–2 **omnes...qui erant senatores** Cf. *Har. Resp.* 13 *cum omnes pontifices qui erant huius ordinis adessent.* The names of the entire College are given in § 12 of the same speech, with the exception of the Pontifex Maximus Caesar (absent in Gaul) and Clodius' brother-in-law L. Pinarius Natta (not a Senator), who probably took no part in the case. *omnes* leaves the absent Caesar out of the reckoning.

2 **Marcellinus** Cn. Cornelius P. f. Lentulus Marcellinus spoke first as Consul-Designate. His father, by birth a plebeian Claudius Marcellus, had become a Lentulus by adoption (cf. Münzer, *RE.* IV. 1390. 44), so that the adoptive surname Marcellinus was hereditary in his case as in that of Lentulus Clodianus (cf. 19 (I. 19). 2 n.). He passed it on to his son and grandson. His enmity with Clodius was of long standing, for he was almost certainly one of the three Lentuli who prosecuted him for sacrilege in 61 (Schol. Bob. (Stangl), p. 89). Though a former Legate of Pompey's against the pirates, he was a staunch opponent of the 'Triumvirs' during his Consulship and resisted the wishes both of Pompey and of his brother (?) Lentulus Spinther in the matter of Ptolemy Auletes' restoration (cf. *Fam.* I. I. 2, II. 2. 1), gaining a warm testimonial from C. (*Q.Fr.* II. 5. 2). Thereafter he disappears.

3–4 **quid essent...secuti** So often, e.g. *Planc.* 39 *remota ratione illa quam in decernendo secuti sumus*; *Div. in Caec.* 10 *ut in constituendo accusatore quid sequi possitis habeatis*; *Verr.* II. I. 72 *secutum id esse Neronem et eius consilium* sqq.

4 **M. Lucullus** Cf. 18 (I. 18). 3 n. Servilius Isauricus was his senior in election to the College as well as to the Consulship (cf. *Har. Resp.* 12). But the Consulars spoke in an order predetermined by one of the Consuls: cf. 13 (I. 13). 2 n. (*rogatum*).

9–10 **neque...factus** Cf. *Tull.* 6 *nox tibi finem dicendi fecit*; Ateius Capito *ap.* Gell. IV. 10. 8 *cum finem non faceret.* No need for *esset* (Otto). The statement, literally untrue, is as natural as 'he went on interminably' would be in English.

11 **odio et strepitu** Almost a hendiadys, 'exasperated outcry': cf. *Q.Fr.* III. I. 24 *concursu magno et odio universi populi paene adflictus est.* Clodius was *odiosus in dicendo* like the bad speaker of *Mur.* 30: cf. Fronto, *ad M. Caes.* IV. 5. 3 *causidicali odio et taedio.*

13 **Serranus** Sex. Atilius Serranus Gavianus, originally of obscure family but adopted into an ancient consular house (cf. *Sest.* 72), was Quaestor in C.'s Consulship and originally well disposed to him (cf. *Cl.Rev.* N.S. 12 (1962), p. 195). He changed sides, allegedly for money, early in his Tribunate. For his conduct on 1 January cf. 71 (III. 26) n.

19 **Cornicinus** Serranus' father-in-law, Cn. Oppius Cn. f. Cornicinus. He served on Pompeius Strabo's staff in 89 (*C.I.L.* I (2). 709). The cognomen suggests that he claimed descent from Sp. Oppius Cornicen the Decemvir.

22 **ei...concessum est** The conclusion is hardly dramatic enough to support an ethic dative such as C. has in 14 (I. 14). 5 *hic tibi rostra Cato advolat* and elsewhere. *ei* (from t^i) is the easiest substitute, *illi* and others less likely have been suggested.

5, 2 **illam** Sc. *Clodianam.*

4 **consili** Cf. Mommsen, *St.* I, pp. 311 f.

5 **Tusculanam sqq.** We might have expected *Tusculanam villam HS quingentis milibus, Formianam ducentis* sqq. (Wesenberg).

9 **illi quidem** The Consuls and their *consilium.*

11 **nam hoc...profuisset** Excellent sense. If C.'s delicacy had counted either way, it would have counted in his favour: cf. Quint. *Decl.* CCLIV (Ritter, p. 41. 7) *ut verecundia ei adeo nihil apud vos prosit ut etiam in causa plurimum noceat.* For *id...hoc* cf. 176 (IX. 9). 3 *ineptum id quidem, sed, puto, hoc* (Δ: *id* Σ) *simulat*; Q.Fr. I. I. 7 *id autem sit magnum et difficile ceteris, sicut est difficillimum: tibi et fuit hoc semper facillimum et vero esse debuit.* Just. XIV. 4. 2 presents a similar sort of sentence: *cernitis...habitum atque ornamenta ducis vestri, quae mihi non hostium quisquam imposuit; nam hoc etiam solatio foret* (cf. *de Orat.* II. 192 *nam id esset levius*). Of the clumsy distortions sometimes preferred the less said the better.

11–12 **mi T. Pomponi** Cf. 67 (III. 22). 3 n.

13 **pinnas** Cf. Hor. *Ep.* II. 2. 50 *decisis humilem pinnis.*

15 **interventu** Apparently Varro was visiting Atticus at Buthrotum.

6, 1 **reliqua** I.e. *de reliquo tempore.*

2 **a Pompeio** Cf. 367 (XIV. 13). 4 *legari a Caesare.* Pompey nominated his Legates to the number granted by the Senate (73 (IV. 1). 7); cf. 90 (IV. 15). 7 n.

5 **prope omnium fanorum, lucorum** Clearly no gloss. *fanorum, lucorum* has a formal sound, as of standard terminology in appointments to *legationes votivae* (cf. 24 (II. 4). 2 n.). *prope omnium* seems to mean that the longer away the better, *prope* being used much as in Hor. *Sat.* II. 3. 32 *insanis et tu stultique prope omnes*, simply to soften a sweeping statement, somewhat at the expense of logic. *pro P.R.* (Orelli) and *prope⟨diem⟩* (Constans) carry no conviction.

6 **nostrae...meae** 'Die von uns angestellten Erwägungen des mir Angemessenen', Sternkopf: cf. *Q.Fr.* III. 6. 1 *consilium nostrum...profectionis tuae.* A rather cumbrous phrase, particularly as *nostrae rationes* might naturally be taken as 'my interests', as in *Fam.* II. 18. 2 *de rationibus tuis cogitanti* and commonly; nor does it seem very likely that Atticus and C. would

have concerted such a plan before the latter's return to Italy. But Lambinus' witness as to Z is against the theory of a gloss, and, as Sjögren pointed out (*Comm. Tull.* p. 158), a glossator would have been more likely to write *meae utilitates*. It is just possible that C. himself wrote *utilitates meae* (cf. *Deiot.* 13 *utilitatibus tuis paruit*) absent-mindedly, having both phrases in his head; a slight interruption could have been responsible.

7, 2 **scis** I.e. 'you can imagine'.

4–5 **etsi...careo** In all probability *non* was in the archetype; small omissions in M are common. Nor is C. likely to have said that he could easily do without a place near the capital. But if *non* is right, *etsi* is indispensable; C. had no *suburbanum* other than at Tusculum. In fact the Tusculanum was not sold, though in January 56 C. was thinking of acquiring another villa in its place: cf. *Q.Fr.* II. 2. 1 *sed ego adhuc emi nihil. Culleonis auctio facta est. Tusculano emptor nemo fuit; si condicio valde bona fuerit, fortassis non amittam* (where *Tusculano* is surely Culleo's villa, not C.'s).

6 **in ea re** This must surely refer to the hire of roughs to resist Clodius during C.'s exile: cf. 68 (III. 23). 5 n. Milo's bands in 57 (cf. *Sest.* 86, etc.) may well have been paid for by C.'s friends.

6–7 **tu absens, ⟨tui⟩ praesentes** Cf. *Fam.* I. 5*a*. 3 *me tibi absenti tuisque praesentibus cumulate satisfacturum*, XI. 27. 4 *quod studium vel in absentem me vel in praesentis meos defuit?*, XI. 29. 2 *et me absentem et meos praesentis*; *ad Brut.* 18. 3 *neque ego illum absentem monere destiti nec accusare praesentis eius necessarios*; also *Fam.* I. I. 4, where I read *nostram fidem omnes, amorem tui absens tu* (*absentis* codd.), *praesentes tui cognoscent* (cf. *Philol.* 105 (1961), p. 72). As in such antitheses the persons regularly correspond, I have preferred *tui* to *mei* or *nostri*. Atticus will have shown C. his letters from Rome when they were together in Epirus during the last months of the exile (cf. 72 (III. 27) n. (*cito videbo*)). *nos* (Madvig) makes *in ea re* refer to the recent past, since C.'s return; but the expenses would have been incurred earlier.

8 **defensores** Like Sestius and Milo.

10 **μυστικώτερα** 'De uxore significat' (Manutius), as the pointed omission of her name in the next sentence shows.

75 (IV. 3)

I, I **quid...ea** 'Constructio Plautina (cfr. Trin. 1168, Poen. 542) neque ab usu Tulliano, ut videtur, aliena, cfr. Madvigius ad II de fin. 7[22]' (Sjögren). *mea* does not suit the contents of the letter.

4 **sed velim** Sjögren prefers *verum ut* (Manutius), others *sed ut* (ς).

5 **sensus** 'Sentiments', 'way of thinking': cf. 377 (XV. 1). 3 *omnem eius sensum*; *Fam.* I. 9. 17 *non offendes eundem bonorum sensum.*

2, 1 **hominibus** They are viewed merely as instruments: cf. K.–S. I, p. 380.

2 **porticus** I.e. as much of it as had been reconstructed. The rebuilding was still going on (*reficiebatur*).

5 **urbe** Cf. *Cael.* 78 (apparently of the same incident) *Sex. Cloelius... qui in Palatio atque in urbis oculis servitia ad caedem et ad inflammandam urbem incitavit.*

7 **nulli** Cf. K.–S. II, pp. 521 f.

7–8 **vel ante demens** *demens* (cf. *Phil.* III. 31 *nec ruere demens nec furere desinit*; *Dom.* 3 and *Har. Resp.* 51 *ille demens* (i.e. Clodius)) cannot be the whole answer. *post hunc vero furorem* needs for balance a reference to the past (cf. *etenim antea* followed by *post has ruinas* below), otherwise *vero* has no function (cf. e.g. *Fam.* XIII. 24. 1 *cum antea...tum vero, postea quam* sqq.). It is of course possible that the lacuna is more extensive.

10 **nolebat** Milo had already made one fruitless attempt to bring Clodius to trial early in 57 (cf. A. C. Clark on *Mil.* 39 *in iudicium bis... vocavit*), and was about to make another. I see no great difficulty in *nolebat* (virtually = *detrectabat*). *tollebat* (Manutius) is no doubt possible (cf. *Sull.* 15 *ille ambitus iudicium tollere ac disturbare...voluit*), but the imperfect would be rather odd.

14 **†Decimum** The man is also mentioned in *Dom.* 50 *Decimis et Cloeliis auctoribus* as one of Clodius' lieutenants, but the name is more than suspicious. It could hardly be anything but a praenomen at this period; but so obscure a person would not be named by praenomen alone, unless with an ironic familiarity which does not belong in either passage. In view of the double corruption the real name must have been very similar, probably either *Decimium* (Orelli; but certainly not the Decimius of 89 (IV. 16). 9) or *Decium*, in the latter case perhaps identical with Antony's supporter in *Phil.* XI. 13 and XIII. 27.

dissignatorem On the forms here and in 78 (IV. 4*a*). 1 *dissignationem* cf. *Thes.* s.vv. Pollock in *RE.* V. 1199 distinguishes two sorts of *dissignatores*, subordinate officials who acted as ushers in the theatre (and probably elsewhere) and undertakers who arranged funeral processions. The two functions could presumably be combined. *dissignatores* were debarred under a law of Caesar's in 45 (not now usually thought to be the lex Iulia Municipalis) from certain civic offices: cf. *Tab. Herac.* 94 (Bruns, *Font. Iur. Rom.*[7] p. 107) compared with *Fam.* VI. 18. 1.

15 **Gellium** Henchman of Clodius, known mainly from C.'s onslaught in *Sest.* 110 f. Himself a knight, stepbrother to L. Marcius Philippus, Cos. 56, he was almost certainly brother of L. Gellius Poplicola, Cos. 72 (cf. Münzer, *RE.* VII. 992), and the elder of the two Gellii lampooned by Catullus (74, 80, 88. 3, 89. 3). Münzer's at first sight startling conjecture

that he is to be identified with Atticus' friend Gellius Canus (making his name Q. Gellius L. f. Poplicola Canus) has a good deal in its favour (cf. 302 (XIII. 31). 4 n.).

3, 3 **Tetti Damionis** From his cognomen perhaps, but by no means certainly, a freedman or descended from one. Clodius had a freedman called Damio (Ascon. 47. 1).

5 **diaeta** Cf. Cels. praef. 9 *iisdemque temporibus in tres partes medicina diducta est...primam* Διαιτητικήν, *secundam* Φαρμακευτικήν, *tertiam* Χειρουργίαν *Graeci nominarunt*. For *curare* cf. *Thes*. III. 1503. 65. C. is particularly fond of medical imagery: cf. the elaborate specimen in 115 (VI. 1). 2.

7 **trudi** Cf. *Tusc*. I. 71 *ad mortem trudi*.

Acidinos The Acidini were a branch of the gens Manlia, prominent in the Punic Wars and the decades following (it may have still survived: cf. 271 (XII. 32). 3): cf. *Leg. Agr*. II. 64 *tum cum haberet haec res publica Luscinos, Calatinos, Acidinos, homines non solum honoribus populi rebusque gestis verum etiam patientia paupertatis ornatos*. In both passages an individual is meant, L. Manlius Acidinus Fulvianus, Cos. 179, of whom Scipio Maluginensis said *virum bonum egregiumque civem esse arbitror* (*de Orat*. II. 260).

postea Not here, I think, = *in posterum*.

8 **Milonis** T. Annius T. f. Milo Papianus became an Annius through adoption by his maternal grandfather (Ascon. 53. 13). As Tribune in 57 he took a leading part in promoting C.'s recall, countering Clodius' gangster tactics by successful imitation. Their warfare ended only with Clodius' murder in 52, followed by Milo's exile in Massilia. He died in 48, attempting with Caelius Rufus to raise Italy against Caesar.

9 **hora quinta** On 12 November 57 (= Julian 24 October) the fifth hour ended just after 11 a.m.

11 **P. Sullae** P. Cornelius Sulla, kinsman of the Dictator and beneficiary in his proscriptions, was elected to the Consulship in 66 along with Autronius Paetus; both were condemned *de ambitu* and lost the prospect of office. Sulla retired to Naples. Prosecuted again in 62 for complicity in Catiline's plot, he was successfully defended by C., according to Gell. XII. 12. 2 ff. in return for a large sum (cf. *Invect. in Cic*. 3). He was now back in Rome (cf. 92 (IV. 18). 3 n.). He is generally supposed to have been militarily active in the Civil War and to have commanded Caesar's right wing at Pharsalia (cf. Caes. *B.C*. III. 51. 1–5, 89. 2; App. *B.C*. II. 76; 236 (XI. 21). 2); but I suspect that Caesar's lieutenant was his homonymous son (cf. 236 (XI. 21). 2 n.). In 48 the elder Sulla would have been at least 60, and C.'s account of him at his trial does not suggest military capacities. The son was *puer* in 62 (*Sull*. 88) and could therefore have been 30 or even older (cf. Marquardt–Mau, *Privatleben*, pp. 127 ff.) in 48, perhaps Quaestor. Caesar liked young commanders; Antony was 32 at the outbreak of war, Curio about 34.

Be this as it may, the father was a prominent personage (*Fam.* XV. 17. 2 πρόσωπον πόλεως) after the war, though further profiteering under Caesar did not improve his reputation (*Off.* II. 29, etc.). C. pretends to no compunction in recording his death in 46 (*Fam. l.c.*).

12 **ex Anniana Milonis domo** I.e. the house which Milo inherited from T. Annius, on the Clivus Capitolinus (*Mil.* 64) just across the valley from the Cermalus. *Milonis* (del. Cobet), or for that matter *Milonis domo*, could be spared, but C. is anxious to make his account as clear as possible.

12–13 **Q. Flaccus** Otherwise unknown. He probably came of the ancient noble house of the Fulvii Flacci, perhaps descending from Servius (Cos. 135) or Gaius (Cos. 134). The Valerii Flacci did not use the praenomen Quintus; while mention by praenomen and cognomen suggests a *nobilis*: cf. R. Syme, *Historia*, 7 (1958), pp. 172 f. The other leading Fulvian family, the Nobiliores, was also languishing in the shadows at this period (cf. 92 (IV. 18). 3 n.).

14 **sed ille sqq.** *sed ille se in interiora* (Orelli) *aedium Sullae abdidit* (ς) must come near the sense, but the wording seems irretrievable. Perhaps it contained something more recondite, as *in interiorem* ⟨μυχόν⟩ (or *mucum*: cf. Marx on Lucil. 1075). Orelli's ... *Sullae. exin senatus* appears in Sjögren's text among others, but *exin* should not be introduced by conjecture into Ciceronian prose: cf. *Thes.* V (ii). 1506. 69.

17 **familiari tuo** Probably Hortensius: cf. 45 (II. 25). 1. But I do not believe that *verissimae* is ironical; irony would be much less than courteous to Atticus, who may on the other hand, friend of Hortensius though he was, have reflected on his consistency—perhaps answering 74 (IV. 2). 5. One is tempted to suggest Pompey (cf. 13 (I. 13). 4 *tuus autem ille amicus*), but he had probably left Rome on his corn-gathering expedition by now (cf. Plut. *Pomp.* 50).

18 **constantia vitae** A common expression in C.: cf. *Sull.* 73; *Acad.* I. 7, etc.

furere Cf. 220 (XI. 9). 2 *hic Ligurius furere.*

19 **comitia sua** Clodius was standing for the Curule Aedileship. These elections, which would normally have taken place in July, had been held up. Clodius was in fact elected in the following January (*Q.Fr.* II. 2. 2; cf. Mommsen, *St.* I, p. 584, n. 5).

Milo Obviously his name, not *Sestius* (Manutius), must be added in view of the sequel (§4).

20 **de scripto** This was unusual: cf. *Planc.* 74 *recitetur oratio, quae propter rei magnitudinem dicta de scripto est.*

23 **dies comitialis** Apart from the Nundinae the remaining days of November were all *comitiales* by the calendar, but no elections could be held during the Plebeian Games, which lasted until the 17th: cf. Mommsen, *St.* III, p. 1056, n. 2.

23 **de caelo servaturum** Cf. 36 (II. 16). 2 n. This seems to be the first recorded exercise by a Tribune of the right of *spectio*: cf. J. Bleicken, *Das Volkstribunat der klassischen Republik* (*Zetemata*, 13 (1955)), p. 26, n. 1. Possibly Clodius gave Tribunes this right in 58. It is at any rate clear that he did not abolish the right of *spectio* altogether, but the extent and manner of his interference, amounting to the repeal of the *lex Aelia et Fufia* as C.'s language in *Pis.* 9 and elsewhere suggests, is in dispute: see R. G. M. Nisbet *ad loc.*

4, 1 **turbulentae sqq.** The plurals can hardly be taken strictly, since it is most unlikely that each of the three made more than one *contio*; probably each made one.

2–3 **haec...futura** Usually misunderstood, as e.g. by Winstedt: 'But the end of it all was that the elections would have taken place, if Milo had not reported', etc. (Wesenberg might well want to add *erant*). *summa* is 'what the speeches came to' (apart from sound and fury). Metellus and the others made it clear that no notice would be taken of Milo's declaration of intention unless he actually 'obnuntiated' on the Campus Martius where the voting took place.

4 **ante mediam noctem** I.e. on the night of the 18th (see below on § 5 *VIIII Kal.*). C. writes *XII Kal.* because he is thinking of the day's proceedings. This expression, not noticed by Unger in *Philol.* 51 (1892), pp. 212 ff., makes against his view that the Roman day was always held to start at midnight (cf. 91 (IV. 17). 4 n.).

8 **fratrum** The two Claudii were the sons of Metellus Nepos' aunt, Caecilia Metella.

11–12 **de nocte** 'Before daybreak'; cf. *Mur.* 69 *prope de nocte.*

13 **†prope deviis** 'Almost a detour' is meaningless. A route is devious, as in 364 (XIV. 10). 1 *ut Trebonius itineribus deviis proficisceretur in provinciam* and elsewhere, or it is direct. Perhaps read *prope⟨rans⟩*. For the loss of the final syllable cf. 12 (I. 12). 1 *ad Teren⟨tiam⟩* (*ad te rem*); 60 (III. 15). 4 *quae⟨ro⟩*; 87 (IV. 13). 1 *val⟨de⟩*; 206 (X. 14). 3 *adhiben⟨dus⟩* (*ad idem*); 212 (XI. 2). 2 *cul⟨pa⟩* (*cui*); 218 (XI. 7). 4 *per⟨ge⟩*.

13 **inter lucos** Romulus' asylum, situated in the lower ground between the two heights of the Capitoline Hill, was called *inter duos lucos* (Wesenberg wished to insert *duos* here). Constans (II, p. 89 n.) argues that this cannot be intended because (*a*) Metellus would have had to get there by the Clivus Capitolinus from the Forum, and so would have been visible to Milo; (*b*) *obnuntiatio* probably could only be made at the place of election; (*c*) C. says that it was made in the Campus Martius (*nisi Milo in campo obnuntiasset*). He therefore concludes that woods in the Campus Martius are meant, and that Metellus proposed to hold the election there instead of in the *Saepta*. But his third argument is based on a misunderstanding of the

text (see above). The second seems to be an unjustified assumption. True, Metellus had said earlier that he would only recognize an *obnuntiatio* delivered in the Campus itself, but he modified this later (*postulat ut sibi postero die in foro obnuntietur*). As for the first, I hardly think we can be sure how Metellus, starting from a point unknown, might have gained the Asylum. On Constans' own theory and its drawbacks see W. S. Watt, *Cl. Quart.* 43 (1949), p. 12.

15 **contio biduo nulla** Cf. 14 (I. 14). 1 n.; Mommsen, *St.* I, p. 199, n. 3. The 21st was a regular market-day. As for the 22nd Constans points out that calendars subsequent to Caesar's reform give 18–20 November as *mercatus*, and conjectures that in republican times this annual market lasted for two days, beginning in 57 with the regular *nundinae*, the 21st. If so C.'s wording is rather odd. Anyhow he only says that no *contio* was in fact called.

5, 1 **VIIII Kal.** This has better authority than *VIII Kal.* and fits the context, since C. would not leave a day (the 23rd) unaccounted for. Arguing to the contrary, W. S. Watt (*Cl. Quart.* 43 (1949), p. 13) assumes that *a. d. VIIII Kal. hora noctis nona* means between 2.15 and 3.25 a.m. on 22 November. But in such expressions the hours of the night were regarded as belonging to the previous day: cf. Bilfinger, *Der bürgerliche Tag*, pp. 208 ff. (not, I think, refuted by Unger: see above on §4 and on 91 (IV. 17). 4).

2 **Marcellus** Probably C. Claudius C. f. Marcellus, Cos. 50 (cf. Münzer, *RE.* III. 2734. 36), whose great-grandfather was grandson to the conqueror of Syracuse. C. mentions him and his father (Pr. 80) affectionately in *Sull.* 19 and remained on good terms with both: see the formal letters in *Fam.* XV. His opposition to Caesar as Consul was followed by unheroic neutrality in the Civil War. His son by Octavia was Augustus' nephew and destined successor. Probably he was the Marcellus who, like several other noblemen, asked Atticus to write his family history (cf. Nep. *Att.* 18. 4).

stertebat Evidently *he* did not expect any election on the 23rd.

4 **pauci pannosi** Sc. *sunt* (or rather *esse nuntiantur*). The nominative is not remarkable: cf. 87 (IV. 13). 1 *Milonis nuptiae*; 187 (IX. 18). 2 *sex legiones* et sim. Ablatives would be more so, to say the least. Theoretically *pauci pannosi* is in contradiction to *vacuum sane*, 'quite empty', unless indeed there is a difference between *vacuum sane*, 'empty, really' (i.e. 'as good as empty') and *sane vacuum*; but such things easily occur in talk and familiar writing.

sine lanterna 'If any ragamuffins were abroad before "the ninth hour of night" on that morning, they would need a lantern, and the fact that they had none might well be mentioned by C. to reinforce *pannosi*, "ut eorum ignobilitas et infima condicio notetur" (Manutius).' So Watt (*Cl. Quart.* 43 (1949), p. 17), rejecting the vulgate *linea lanterna* (Bosius). There is no real evidence for canvas lanterns in antiquity (*l.c.* pp. 15 ff.).

illi Clodius and his supporters (not, of course, the *pannosi*).

8 **iam** Implies apparently 'after all that has happened'. With *obviam* (Victorius) C. would probably have written *tulerit.*

11 **invidi et perfidi** Both terms hint at Hortensius. Some have thought that *perfidi* means Pompey, but C. was more apt to remember what Pompey did for him in 57 than what he failed to do in 58.

12 **nobilitati** The *piscinarii* in general: cf. *Balb.* 51 *nobilitatis inertiae*; *Q.Fr.* II. 3. 4 *nobilitate inimica* et sim. *simplex nobilitas* in Ov. *Fast.* II. 286 means 'you guileless nobles'. *nobili* suggests an individual; but who in particular? Cf. 82 (IV. 8*a*). 2 *quod viri boni nusquam.*

6, 3 pro facultatibus nostris 'Quantum nostrae facultates patiuntur' (Manutius)—and no more. *ne...essem* explains why C. could not repay Quintus in full. The sentence has been much misunderstood: cf. W. S. Watt, *l.c.*, pp. 17 ff. Watt himself takes *subsidiis* as dative and (reading *respondemus*) understands that C. had started to repay his brother, in so far as his means allowed, but when Quintus refused to take the money had begun to pay his other creditors. But, apart from the obscurity of the expression, why should C. make the qualification *pro facultatibus...essem* if Quintus did not take the money? It is true that in 74 (IV. 2). 7 C. speaks of his friends' (financial) kindness as having been exhausted earlier on; but that was nearly two months previously and in a different connection. Further loans in the interval are not ruled out.

4 **respondemus** *respondimus* (Klotz) is usually and perhaps rightly read. But *essem* may be taken as the apodosis of an implied condition, 'if I repaid him in full and at once'. The verb is often used of paying debts: cf. e.g. 84 (IV. 10). 2; 412 (XVI. 2). 2.

5 **statu** C.'s political position *vis-à-vis* Pompey seems to be especially in his mind.

76 (IV. 4)

4 **hoc ipsum** C.'s pleasure at Cincius' news has already been implied; hence *ipsum.*

8 **sis** Cf. 93 (IV. 19). 2 *quo die venies tu...apud me cum tuis maneas.* The second personal jussive present subjunctive is common in Comedy. C. hardly uses it outside his letters and seldom in them; but the evidence appears sufficient: cf. K.–S. I, p. 186.

77 (IV. 7)

On the dating of this and the following letters see Appendix II.

I, I sollicitum Cf. Q.*Fr.* II. 6. 2 of 9 April: *a. d. VIII Id. Apr. sponsalia Crassipedi praebui; huic convivio puer optimus, Quintus tuus meusque, quod perleviter commotus fuerat, defuit. a. d. VI Id. Apr. veni ad Quintum eumque* *vidi plane integrum, multumque is mecum sermonem habuit et perhumanum de discordiis mulierum nostrarum. quid quaeris? nihil festivius.* Quintus junior was now ten years old!

3 **Chaerippus** A friend, perhaps a freedman or client of Q. Cicero, who had been with him in Asia (*Q.Fr.* I. 1. 14); scarcely to be identified with the friend of Q. Cornificius mentioned in *Fam.* XII. 22 *a*. 4 and XII. 30. 3.

mera monstra Cf. 178 (IX. 11). 4 *monstra narrant*; *Q.Fr.* II. 5. 3 *monstra promulgarunt*; 180 (IX. 13). 1 *mera scelera loquuntur*; 178 (IX. 11). 3 *sermones minacis,...meras proscriptiones, meros Sullas* et sim.

4 **Apollonio** Otherwise, like Terentius, unknown. The latter must have been a bankrupt Knight. It was a piece of impudence for a mere Greek to claim the privilege of going bankrupt.

illi Sc. *male faciant* vel sim.: cf. 397 (XV. 20). 3 *di illi mortuo qui umquam Buthrotum!*; 409 (XVI. 1). 1 *di hercule istis!*; 420 (XVI. 11). 5 *di istis! qui = ut, utinam*, elsewhere only pre-classical: cf. K.–S. I, p. 183.

2, 1 **Metello** Possibly the M. Metellus who gave a gladiator show in 60: cf. 21 (II. 1). 1 n. Drumann (II, p. 46, n. 1; III, p. 630, n. 3) has a tenuous hypothesis that he was a son of Metellus Creticus who had supported Herennius' efforts on Clodius' behalf in 60.

οὐχ ὁσίη φθιμένοισιν ἐπ' ἀνδράσιν εὐχετάασθαι (*Od.* XXII. 412). So Pliny, *Ep.* IX. 1. 3 *vitabis illud* οὐχ ὁσίη φθιμένοισιν. Homer's text has κταμένοισιν.

2–3 **cui equidem...** Sc. *maledicam libentius* vel sim. The dative is helped by φθιμένοισιν, but it is of course possible that the text is defective. Cf. however 278 (XII. 38). 2 *o hominem cavendum! quamquam mihi quidem — sed tenendus dolor est.* Editors reading *qui quidem* (Malaespina) understand that 'some unfavourable criticism is suppressed'; but that violates Ciceronian usage. After a relative pronoun *quidem* emphasizes the relative clause, which is sometimes restrictive (e.g. 125 (VII. 2). 3 *epistulas...quae quidem erant tua manu*) or explanatory (e.g. 36 (II. 16). 2 *qui quidem etiam istuc adduci potuerit*), but usually adds a new and important point. The main clause has to be complete in itself—*civis nemo erat mortuus qui quidem peior fuit* could never have been C.'s writing.

3 **nummi** The common assumption that Metellus owed Atticus money is irreconcilable with the context. *thecam non retexeris*, which can only signify the putting of money in or the taking of money out, will then mean 'you won't get your money back', since there would be no question of Atticus' having anything to pay on a bad debt. But this discouraging prophecy nullifies the preceding assurance that Atticus' money is safe. It therefore becomes necessary to understand the debt as due, not from Metellus to Atticus, but the other way round. The circumstances, which were doubtless explained in Atticus' letter if C. did not know them already,

may be imagined somewhat as follows. Metellus had a claim on Atticus, perhaps based on some legal quibble, which he himself had not cared to take into court. Now that he is dead, Atticus is afraid that his heir may pursue the matter. C. reassures him, on the ground that Metellus' heir (unless it were P. Clodius, which, he implies, is out of the question) could not be a greater scoundrel than Metellus himself.

3 **meo periculo** Cf. *Propertiana*, p. 44.

4–5 **†verum sqq.** The sense can hardly be far from Müller's reconstruction *virum fecit non improbiorem quam fuit ipse*, but C. would surely have written *hominem* or no noun at all, not *virum*. Perhaps he wrote *nimirum*. None of the other suggestions (e.g. in Sjögren's text *quid enim vereris, quemcumque heredem fecit, nisi Publium fecit? verum fecit non improbe, quamquam fuit ipse*) even looks like holding water.

3, 1 **praesidia** Against further interference from Clodian gangs.

2 **Laterio** Laterium seems to have been a village (cf. *Q.Fr.* III. 1. 5 *Lateri*) between Arpinum and Anagnia, but Q. Cicero's estate there is called *Laterium* (*ibid.* 5 *in Laterio*; 190 (X. 1). 1 *in Laterium fratris*), never *Laterianum*. Perhaps the village was included in the property: cf. 9 (I. 4). 3 n. (*vicos*). Quintus' offence which roused the Arpinates may have concerned the diversion of a watercourse: cf. *Q.Fr.* III. 1. 4.

3 **ὁ δὲ κ.τ.λ.** *Od.* XVII. 488.

4 **quid superest? etiam** Cf. 126 (VII. 3). 12. There can be little doubt that this is the right reading. *quod superest, etiam puerum* sqq. makes *etiam* otiose, since *puerum Ciceronem* (cf. 85 (IV. 9). 2) here plainly refers to Quintus, not Marcus. *quod superest etiam, puerum* (F. Walter, *Berl. Phil. Woch.* 49 (1929), p. 955) is hardly borne out by *accedit etiam* in *Fam.* V. 12. 8 and *Har. Resp.* 6.

78 (IV. 4*a*)

C.'s efforts to resume his former role in politics had proved a mortifying failure, earning no thanks from the optimates and a stern warning from Pompey (cf. *Fam.* I. 9. 9). Hence his 'palinode', probably the speech *de Provinciis Consularibus* (cf. App. II, p. 233) in late May or June 56, in which he virtually forswore his renewed opposition to the dynasts. Shortly afterwards, as in 59, he retired to his house in Antium, and occupied himself with the arrangement of what was left of his library: cf. Gelzer, *RE.* VII A. 943.13.

1, 1 **dissignationem** Apparently the correct form, though *designatio* seems to have supplanted it in Vitruvius, etc.: cf. *Thes.* s.v. and 75 (IV. 3). 2 n.

2 **Tyrannionis** Cf. 26 (II. 6). 1 n.

4 **glutinatoribus** Their job was to glue the loose strips of papyrus edge to edge so as to make a connected length ready to roll round the central stick: cf. Crönert, *Hermes*, 38 (1903), pp. 404 f.

6 **vos Graeci** Cf. 16 (I. 16). 4 n. (*tui cives*).

7 **σιττύβας** σιλλύβους Graevius; but cf. 79 (IV. 8). 2 and 80 (IV. 5). 3. The word itself should surely be identical in all three places. Hesychius glosses σιττύβαι thus: δερμάτιναι στολαί· τὰ μικρὰ ἱμαντήρια. The second explanation is not far away from tongues of parchment for book labels. As for the lettering, C. would hardly have Latinized here, with *vos Graeci appellatis*, but might well do so in the other two passages.

2, 3 **Piliam** Atticus (now 53) had married her on 12 February of this year (*Q.Fr.* II. 3. 7), his only marriage it would seem. Of her family nothing is known apart from her relationship, whatever it was, to Q. Pilius Celer (cf. 92 (IV. 18). 5 n.). She bore him a daughter in 51 and perhaps died in 44 (cf. 415 (XVI. 7). 8). Cornelius Nepos does not mention her. With C. she always seems to have been on good terms.

adducas This or *adduc* (Orelli) has been assimilated to *adhaerescere* in the MSS. A jussive form is needed because C. is inviting Pilia, not making her visit a condition of Atticus'.

4 **ludum** λόχον Bosius; but then why *gladiatores* (hardly as distinct from their *lanista*)? *ludum* (literally the training school, but the gladiators would naturally go with it) may have become *locum* from *locis* above or *locare* below. Men of rank often owned 'families' or 'schools' of gladiators; as C. Aurelius Scaurus (Val. Max. II. 3. 2), Lentulus Vatia (Flor. III. 20. 3: cf. *Cl. Quart.* N.S. 10 (1960), pp. 258 ff.), Caesar (Suet. *Iul.* 10. 2; Gell. XII. 5. 13), D. Brutus (Vell. II. 58. 2), Caligula (Plin. *N.H.* XI. 144). Cf. Quint. *Decl.* CCLX (Ritter, p. 65. 2) *illos qui ingentes familias occisuri tamen pascunt plausu fovebimus et in omnibus spectaculis adulatione multa prosequemur*. C. Cato had bought some *bestiarii* from Atticus earlier in the year (*Q.Fr.* II. 5. 3).

5 **locare voluisses** Perhaps, as T.–P. suggest, Atticus had not thought them sufficiently trained.

6 **liberasses** Sc. *impensam*: cf. Col. *R.R.* III. 3. 11 *cum sola ea res omnem impensam terreni pretio suo liberet*. The ellipse is not found elsewhere, but may have been business jargon. *se liberare*, 'free oneself from debt', occurs in *Phil.* II. 103 (cf. Sen. *Ep.* 23. 9 *hanc epistulam liberare*), likewise *liber* (*esse*) (Plin. *Ep.* III. 9. 13). But *liber esses* (Pius) is no improvement.

7 **diligenter** Sc. *fac*. He did; cf. 79 (IV. 8). 2 and 80 (IV. 5). 3.

79 (IV. 8)

1, 2 **tyrotarichi** A dish of salt fish and cheese, proverbial poor man's fare: cf. 370 (XIV. 16). 1; *Fam.* IX. 16. 7.

raudusculo C. rather favours this diminutive of *raudus* or *rodus* (cf. Fest. p. 320 (Lindsay) *significat rem rudem et imperfectam*), though only when writing to Atticus. The money may have no connection with that mentioned in 77 (IV. 7). 2, nor is there any reason to relate the *raudusculum* to the *tyrotarichum*.

2 **μήπω κ.τ.λ.** From Sophocles' *Tyro* (Nauck, *Trag. Graec. Frag.* p. 275).

6–7 **Antium...tuum** 'C. wishes to say that Antium is to Rome as Buthrotum to Corcyra, but not having our traditional phrases to express relation, he is bound to have recourse to a rather awkward expression' (T.–P.). *Antium* (Lehmann) after *tuum* hardly improves it. *Corcyrae* is the town, not the island, and, like *Romae*, probably dative rather than genitive: cf. *Verr.* II. 4. 23 *haec civitas isti praedoni ac piratae Siciliensi Phaselis fuit.*

7 **†ειμη κ.τ.λ.** The original apparently had something to do with the proverb οἶκος φίλος, οἶκος ἄριστος (Leutsch–Schneidewin, *Corp. Paroem. Graec.* I, p. 438: cf. 392 (XV. 16*a*)), but none of the restorations proposed is worth repeating. *postea vero* implies a reference to the past (ἀεί μοι φίλος οἶκος?): cf. 75 (IV. 3). 2 n. (*vel ante demens*).

2, 2 **mens** Cf. Virg. *Aen.* VI. 727 *mens agitat molem.*

3 **Dionysi et Menophili** The *librarioli* of 78 (IV. 4*a*). 1. That Atticus' freedman and friend Dionysius (cf. 82 (IV. 8*a*). 1 n.) was sent to C. in this humble capacity seems to me altogether incredible. The name was very common.

4 **sittybae** I cannot see any reason to write (*mi*) *sittybis* (Bosius).

illustrarunt Perhaps with an eye to the literal sense 'brightened'.

5 **vale** *valde* ⟨*laudo*⟩ Lehmann. But cf. with Sjögren 67 (III. 22). 3; 214 (XI. 4*a*).

7 **gesserunt** *gessere* Malaespina. But though C. was willing to tolerate forms in *-ere* (cf. *Orat.* 157) it is more than doubtful whether he ever used them himself: cf. R. G. M. Nisbet on *Pis.* 96. Possibly a word belonging to the next letter has fallen out: e.g. *gesser⟨unt.//her⟩e.*

80 (IV. 5)

1, 1–2 **an...te** I do not think it necessary to delete *an* as a reflection of *a*(*i*)*n* or to replace *tu an* by *tandem* with W. S. Watt (*Rhein. Mus.* 106 (1963), pp. 15 f.). As he points out, interrogative particles in questions after *ain tu?*, though unusual, do occur, and *an* seems appropriate as introducing an expression of dissent 'cum alterius verba liberius referuntur': cf. *Thes.* II. 3. 32 (*ain tu?* can be regarded as in effect such a citation). Similarly in *Tusc.* V. 35 I would read *ain tu? an* (Moser).

1 **mea** 'My writings' like *meorum* in 410 (XVI. 5). 5 and *sua* in 242 (XII. 5). 2. Cf. *Propertiana*, p. 116.

2 **ab eo** Probably Pompey.

3 **exempla duo** *exempla ii* explains the paradosis better than *exemplar*,

a word never used by C. literally of a copy of a document. C. would of course keep one copy himself.

3 **quin etiam** Advocated by Watt (*l.c.* p. 17); *quid ⟨quod⟩ etiam . . .?* had occurred to me. The vulgate *quid? etiam (dudum . . . devorandum est) subturpicula* sqq. leaves *quid* without appropriate function. Evidently something has fallen out. One range of possibilities is sufficiently indicated by the passages which Müller cites *ad loc.*: 13 (I. 13). 6 *novi tibi quidnam scribam? quid? etiam*; 26 (II. 6). 2 (see *ad loc.*); 39 (II. 19). 5 *quid aliud? quid? hoc, opinor*; 92 (IV. 18). 4 *quid aliud novi? etiam*; 126 (VII. 3). 12 *quid superest? etiam* (cf. 77 (IV. 7). 3); Q.Fr. III. 1. 24 *quid praeterea? quid? etiam* (also Plin. *Ep.* III. 14. 6 *quid praeterea novi? quid? nihil*). But in all these cases the writer is casting around for a further item of news to round off his letter. In assembling arguments or excuses such an expression would be exceptional, and surely rather forced. *quid quaeris?*, also suggested by Müller, should precede a summing up rather than an additional point.

3–4 **iam dudum** Cicero uses *dudum* with a present tense only in conjunction with *iam* or *quam*.

5–6 **recta, vera, honesta consilia** The *via optimas* of 20 (I. 20). 3. By itself *vera consilia* could hardly mean 'sound policies'. It is by no means certain that *consilium verum* in 201 (X. 10). 2 (Antony) means a 'sound plan': cf. '*Towards a Text*', p. 50. But *recta* (cf. *Thes.* IV. 451. 5; Liv. I. 27. 1) and *honesta* (cf. *Thes.* IV. 450. 80) are natural terms to use, and *verus* goes naturally with *rectus* and *honestus*: cf. *Fam.* I. 8. 2 *et recta et vera*; *Tusc.* III. 64 *recta, vera, debita,* 73 *rectum quoque et verum*; *Acad.* II. 27 *lex veri rectique*; Sall. *Orat. Macri* 13 *si vera et honesta flagitium superaverit*; *Fin.* I. 25 *recta et honesta quae sint* et sim.

6 **principibus** Cf. 74 (IV. 2). 5. C. harps on the same string in his long apologia to Lentulus Spinther (*Fam.* I. 9) and elsewhere; no doubt with some justice in that the leading nobles were apt to be jealous of his pretensions and took pleasure in provoking him by their cultivation of Clodius, a man of their own class. But his complaints can hardly be more than a smoke-screen for the true reason of his recantation, namely the danger of opposing the coalition just re-established at Luca after Pompey's plain warning (cf. *Fam.* I. 9. 9). C. did not want to go on his travels again.

7 **senseram** Sc. *perfidiam*.

8 **inductus** Cf. 60 (III. 15). 7 *proditus, inductus, coniectus in fraudem*.

8–9 **hoc eram animo** Cf. *Verr.* II. 2. 157 *hoc animo esse Siculos ut . . . arbitrentur. hoc erat in animo* would give a slightly different sense ('I had it in mind that I should', etc.); an infinitive would then be normal.

10 **resipivi** Watt (*l.c.* p. 15, n. 1) points out that this form rather than *resipui* is supported not only by the MSS but by *resipisset* in *Sest.* 80, the only other place where Cicero uses the perfect of this verb.

2, 1 **ea te suasisse** *ea tenus te suasisse qua facerem* (Klotz) is closer to the MSS, but the absence of an object to *facerem* tells against it. Ancient editions have *ea te monuisse, suasisse. tenuisse* seems to have arisen by a sort of dittography; or possibly *monuisse* and *suasisse* were alternatives in the progenitor of Ω: cf. Sjögren, praef. p. xix. *ea tenus te suasisse qua tacerem* (Constans) is approved by Watt; but Atticus appears to have gone further than this: cf. §3 *dices 'vellem iam pridem.'* At any rate he wanted C. to cultivate both Pompey and Caesar privately: cf. 124 (VII. 1). 2 *videsne ut auctore te sim utrumque complexus?*

quae facerem Alluding, it would seem, to the delivery of C.'s pro-Caesarian speeches in the Senate (cf. *Balb.* 61), also perhaps to his private conduct (see above). *facerem* and *scriberem* are here contrasted (cf. 124 (VII. 1). 4 *laudator et factorum et scriptorum meorum*). It must be remembered that the published version of the *de Provinciis* (if that is the 'palinode') may have differed considerably from the original speech.

3 **relabi** Sjögren compares *ad Brut.* 24. 6 *nonne hoc est in easdem tenebras cecidisse*, where however Watt reads *recidisse*.

5 **ἀποθεώσει** ὑποθέσει, another old conjecture printed by many editors, is senseless. ἀποθεώσει is rather extravagant, but perhaps it was facetiously meant—it seems to be quoted from a previous letter, whether of Cicero or of Atticus. It is so close to the MSS that I follow Sjögren, with some misgiving, in accepting it.

scripseram Graphically quite as good as the vulgate *scripsi* (Victorius), and better suited to *fuimus*. Watt (*l.c.* p. 19) points out that *ut scripsi* alone, as a reference to a previous letter, would be unique in Cicero's correspondence (though C. does use it to refer to an earlier passage in the same letter). That objection does not apply to *ut scripseram*: cf. 58 (III. 14). 2 *in Epirum ideo, ut scripseram, non veni*; 420 (XVI. 11). 6 *ego me, ut scripseram, in Pompeianum non abdidi*. There is nothing against the supposition that Cicero had informed Atticus of his intention to publish his 'palinode'; it is indeed much the most likely way for Atticus to have heard about it. But *ut scripsisti* (Watt: better *scripseras*) is also quite possible.

6 **ille** Caesar.

villam At Tusculum (not Antium, as G. McCracken, *Cl. Journ.* 30 (1934), p. 262; C. had no villa at Antium, only a house in the town). C. had acquired this villa by 68 (cf. 1 (I. 5). 7). According to Pliny (*N.H.* XXII. 12) it had formerly belonged to Sulla, some time after 89 since he decorated it with a painting relevant to that year. If *Catuli* is, as generally supposed, the younger Catulus (Cos. 78) he presumably inherited it from his father (Cos. 102) and sold it to Vettius; for the elder Catulus' *Tusculanum* cf. *de Orat.* II. 13. Since the scene of the *de Oratore* is laid in 91, the villa may on this hypothesis have passed into Sulla's hands temporarily, Catulus the

Elder having been (in effect) murdered by the Marians in 87, and been restored by him to the son. Much more probably, *Catuli* is the elder Catulus, and the chain of ownership ran: Catulus, Sulla, Vettius, C. The villa could have been acquired by Vettius (cf. 44 (II. 24). 2 n.) from Sulla, who may have bought it from Catulus in 89–88. That the younger Catulus owned a *Tusculanum* does not necessarily follow from *Acad.* II. 148 *in Tusculanis nostris*, since *nostris* may apply only to the speaker, Lucullus, and the person he addresses, C.; but if he did have one at the time of the dialogue, which was subsequent to C.'s Consulship (cf. *ibid.* 62), it was a different property. —The complaint of the *nobiles* is notably paralleled by what his prosecutor said about Cornelius Balbus, who owned a *Tusculanum* which had once belonged to Metellus Numidicus and L. Crassus (*Balb.* 56). There as here C. makes the point that the villa had also had humbler proprietors. The *pro Balbo* belongs to this same summer, and C. doubtless had his own case in mind when he spoke.

7–8 **me emisse** *me* survives in *etiam* of M as well as in Z^b.

9 **oportere** Assimilated in the MSS to *oportuisse* preceding (*sed* may have helped). The *nobiles* cannot have said that C. ought to have sold his Palatine house before he built it. Commentators explain them as meaning that he ought to have sold the site, but that is not in the Latin: cf. *Rhein. Mus.* 107 (1964), pp. 283 ff. Besides, C. is attacking his critics on the score of inconsistency; but there would be nothing inconsistent or unreasonable in saying that he should not have built the new house and should have sold the site. *oportere* gives a characteristic verbal point: 'they say I ought not to have built the place, but ought to be selling it'—as though, if he had a right to sell the house (now perhaps getting on towards completion after over eight months' work (cf. 74 (IV. 2). 7; *Q.Fr.* II. 6. 3)) he had not had an equal right to build it.

10 **etiam** *et* can hardly stand in C.: cf. *Thes.* V (ii). 909. 21. It could of course simply be omitted (Boot). Watt suggests *ei*.

laetati sunt So to Lentulus, *Fam.* I. 9. 10, 20.

11 **finis sit** Boot proposed ⟨*sed*⟩ *finis sit* comparing *Mil.* 105. Sjögren counters with *Off.* III. 115 *sit iam huius loci finis*. But see translation.

4, 2 **viaticam Crassipes praeripit** Piso having died in 57, Tullia, now nearly 23, was betrothed on 4 April of this year (*Q.Fr.* II. 6. 1) to Furius Crassipes (praenomen unknown), a young man of minor patrician family. The marriage followed within a year or so (cf. *Fam.* I. 9. 20 (December 54) *in mei generi Crassipedis hortis*), but did not last long. In 51 Crassipes was in Bithynia, probably as Quaestor, and there received a recommendatory letter from C. (*Fam.* XIII. 9). Their relations seem to have remained superficially amicable, if cool: cf. 178 (IX. 11). 3; 219 (XI. 8). 2 n. The vulgate *viaticum* is generally thought to mean that the expense of Tullia's dowry had cleaned C. out, so that he had no money to travel with. That is implausible on several counts: (*a*) Tullia's dowry would hardly be paid so soon (cf. 212 (XI. 2). 2), and as she presumably recovered the dowry provided for her first marriage C. ought not to have been heavily mulcted. (*b*) The only travel which C. is known to have had in mind at this time was the return journey from Antium to Rome. It is unlikely that he was still thinking of a *votiva legatio* (cf. 74 (IV. 2). 6), and Constans' hypothesis that C. had had a tour of his villas in view is pure assumption. (*c*) It is hard to imagine C. giving up a trip of the latter kind because of a shortage of funds. The expense of such journeys could not have been formidable since the servants and transport would be his own, and lodging between villas was mostly provided by friends or by his own *deversoria*. The right answer emerges from a correct understanding of the next sentence. Atticus, it is to be supposed, had invited C. to dine with him on the evening of his return to Rome: cf. 76 (IV. 4) *quo die venies, utique cum tuis apud me sis*; 93 (IV. 19). 2 *quo die venies, tu, si me amas, apud me cum tuis maneas*; 238 (XII. 2). 2 *iam te videbo et quidem, ut spero, de via recta ad me.* But instead of dining with Atticus, C. now proposed to dine with Crassipes at the latter's 'gardens' (cf. 81 (IV. 12)). Crassipes was therefore 'forestalling' the 'welcome-home' dinner which Atticus had hoped to provide—though the plans were still not quite definite (*sed viderimus*). For *cena viatica* = *cena adventicia* cf. Plaut. *Bacch.* 94 *ego sorori meae cenam hodie dare volo viaticam.* The ellipse of the substantive is unremarkable (cf. K.–S. II, p. 550); there happens to be a parallel in Petr. 90. 5 *adventicia.*

2–4 **tu...viderimus** Atticus is supposed to object. He might find something odd, even slightly improper, in C. spending the first evening of his return to the capital at a suburban dinner-party. That *horti*, perhaps through Clodia's notorious Tiberside playground (*Cael.* 36, etc.), had come to sound a little 'dashing' in respectable ears is evident from *Cael.* 27 *eum qui nullum convivium renuerit, qui in hortis fuerit, qui unguenta sumpserit, qui Baias viderit*: cf. *Rhein. Mus.* 107 (1964), pp. 279 f. *cogito* (Watt) for *tu* is needless. Understand *inquis* vel sim. In a staccato passage like this a rare ellipse is not surprising; cf. 366 (XIV. 12). 3 *tu si quid erit de ceteris, de Bruto utique quicquid* (sc. *scribe*). But anyone discomposed by the lack of a precise parallel may include *tu* in the 'quotation', comparing 188 (VIII. 9). 2 *num igitur peccamus?* For the imaginary objection followed by asyndetic answer cf. 82 (IV. 8*a*). 3 *dices 'tantulane causa?' permulta ad me detulerat non dubia de Firmanis fratribus.* For *recta* cf. 238 (XII. 2). 2 (see above); *Fin.* V. 96 *et recta quidem ad me* et sim.

3 **quid enim tua?** Cf. Plaut. *Amph.* 1003 *quid ⟨id⟩ mea?* A day more or less before the reunion does not really matter to Atticus.

5 **cum structione et sittybis** *structio* refers to the tiers of bookshelves (*pegmata*). Besides these and the book labels, the *librarioli* also painted the

ceiling and walls, where visible: so Birt, *Rhein. Mus.* 64 (1909), pp. 469 f. *constructione* (or *constrictione* (Hertzberg)) is usually supposed to refer to binding (cf. 78 (IV. 4*a*). 1 *glutinatoribus*), but even if the word could have such a meaning it would consort oddly with *pinxerunt*; see further W. S. Watt, *l.c.* pp. 21 ff.

81 (IV. 12)

1 **Egnatius** L. Egnatius Rufus, with whom both Ciceros had business relations. In a letter of recommendation of doubtful date (*Fam.* XIII. 43) C. writes of him *quo ego uno equite Romano familiarissime utor.*

†Halimeti Possibly, as Corradus conjectured, to be identified with the equally obscure 'Thallumetus' of 105 (V. 12). 2.

2 **Aquilio** Cf. 10 (I. 1). 1 n.

3 **Macroni** Nothing is known of Macro or of what he wanted from C.

4 **esse** *posse* (Lambinus) is added by editors, without necessity: cf. 69 (III. 24). 2 n. (*decernere*).

6 **cena apud me** A revision of the proposal in 80 (IV. 5). 3.

7–8 **quasi in deversorio** Perhaps with a hint of self-excuse (see above). The halt *in hortis* might be regarded as a stage on the journey.

8 **facio fraudem** For the phrase cf. Plaut. *Mil.* 164 *ut ne legi fraudem faciant aleariae*; Plin. *N.H.* X. 140 *in fraudem earum* (sc. *legum sumptuariarum*). I know no evidence to support Manutius' explanation that Senators in Rome were ordered to attend sittings while those not in Rome could not be so ordered (cf. Mommsen, *St.* III, pp. 915 ff.); besides, C. must be referring to a particular decree. Corradus was probably right in thinking of sumptuary restrictions, presumably on expenditure in public places (a *deversorium* might be private or public). Pompey and Crassus, dissuaded by Hortensius, gave up their plans for a sumptuary law in 55 (Dio, XXXIX. 37. 2–4: cf. Kübler, *RE.* IV A. 907 f.).

9 **praesto** In what capacity is uncertain; not at any rate to attend his wedding (cf. 87 (IV. 13). 1). Possibly Clodius' prosecution of Milo *de vi* is relevant. That had been fixed for 7 May (*Q.Fr.* II. 6. 4), but there may have been a postponement.

ibi *domi.*

10 **praemonebo** Much the most satisfactory correction of *promonebo*, a *vox nihili* (*praemoneri* is rightly read by Müller, etc., in *Har. Resp.* 10, *promoveto* should be read with Ritschl (against Leo and Lindsay) in Plaut. *Pers.* 311). C. will send word to confirm the arrangement when he arrives on the Kalends.

82 (IV. 8 *a*)

1, 1 **Apellas** Graphically hardly further from *aperias* than the vulgate *Apenas* (ς). The one name is common, the other unattested. M. Fabius Gallus' freedman mentioned in *Fam.* VII. 25. 2 could be the same man.

2 **legem** We do not know what law or who proposed it.

4 **ludis** Probably the *ludi plebeii*, which normally ran from 4 to 17 November: cf. Appendix II, p. 234. This sentence is not a separate item thrust into the surrounding political comment; it tells Atticus that C. would be staying where he was, in connection with the previous injunction to write at once.

5 **Dionysio** M. Pomponius Dionysius (on the praenomen cf. 90 (IV. 15). 1), a learned freedman of Atticus' and tutor to the young Ciceros (cf. *ibid.* 10). Much is heard of him in these letters, nothing otherwise.

2, 1 **Trebonio** Atticus seems to have expressed some opinion about his forthcoming Tribunate, in which he passed his law assigning provinces to Pompey and Crassus. C. Trebonius C. f., whose father, *civis acerrimus* (*Fam.* X. 28. 1), was a distinguished Roman Knight (*Phil.* XIII. 23), had in some way supported C. as Quaestor (cf. *Fam.* XV. 21. 2); but down to 44 he figures as a steady 'triumviral' or Caesarian partisan. After his Tribunate at any rate he followed Caesar, under whom he served as Legate in the Gallic and Civil Wars, who made him Praetor in 48 and Consul Suffectus in 45, whose murder he conspired and was the first to expiate. He and C. were always on very friendly terms: see the letters between them, *Fam.* X. 28, XII. 16, XV. 20, 21.

Domitio Foiled in his candidature for the Consulship of 55 (see below). He succeeded next year.

2 **σύκῳ κ.τ.λ.** An unplaced fragment of New Comedy (Kock, III, p. 445).

4 **ab iisdem** Sc. *laeditur* vel sim. The dynasts are meant.

5 **nusquam** Sc. *apparuerunt* rather than *fuerunt*. *nusquam esse* usually means 'not to exist'.

6 **merito** Why? C. can hardly mean simply that Domitius 'tempted his fate' by threatening to deprive Caesar of his army if elected (cf. Suet. *Iul.* 24. 1); his own rejection of Caesar's offers in 59 would surely have occurred to his mind. Probably he disapproved of some other action or actions of Domitius, of whom, despite a debt of gratitude for support in 65–64, he had a very poor opinion: cf. 10 (I. 1). 3 n.

7 **tot annos** Cf. *Leg. Agr.* II. 100 *qui non in cunabulis sed in campo sunt consules facti*; on Domitius' consular ancestry see 10 (I. 1). 3 n.

8–9 **praesertim cum...petat** See Appendix III.

11 **habeant** Sc. *illi*, the dynasts. With *habeat* we must understand

Pompeius, an awkward change of subject—not intolerably so perhaps, but the correction is minimal. Another possibility is to supply *noster* ($\overline{nr}$) before *non*: cf. 104 (V. 11). 2; 150 (VII. 26). 2.

12 **factorum** Such notebooks presumably contained a list of Consuls for chronological reference, extending back a decade or so.

3, 1 **de Natta** Almost certainly about his death: cf. 83 (IV. 6). 1 *de Lentulo* et sim. L. Pinarius Natta, scion of an immemorially ancient but politically withdrawn patrician house and stepson of L. Licinius Murena (Cos. 62), is commended as a young man of spirit and promise in 63 (*Mur.* 73). Five years later, as a recently elected Pontifex he co-operated with his brother-in-law Clodius in consecrating the site of C.'s house (*Dom.* 117 ff.).

2 **poëmate** Probably *de Temporibus Suis*: cf. *Fam.* I. 9. 23.

effugere Cf. Plin. *Ep.* IX. 25. 3 *tu passerculis et columbulis nostris inter aquilas vestras dabis pennas, si tamen et sibi et tibi placebunt; si tantum sibi, continendos cavea nidove curabis.*

sinam Perhaps there is not much to choose between *sinam* and *sinas* (Manutius), but I prefer the former as more direct, the decision being properly C.'s: cf. 169 (IX. 2 *a*). 1 *quid si hoc ipso premar? accipiam?*

3 **Fabio Lusco** Otherwise unknown. A freedman as Constans thought, comparing 127 (VII. 4). 1 *frugi hominem ac, ne libertinum laudare videar, plane virum bonum.* But C. can describe a *praetorius* as *frugi homo* (cf. *Font.* 40), and the cognomen does not support this hypothesis. But Constans (II, p. 123, n. 1) was surely right in his belief that there is more in this passage than meets the eye. It would be very strange if the *Firmani fratres*, of whom C. thought it so desirable to be kept informed through this obscure personage, were the equally obscure Gavius and an unknown brother. His evident uneasiness suggests, and §4 as good as proves, that high persons were involved, perhaps, as Constans suggests, the forthcoming Consuls Pompey and Crassus, though Pompey's possessions in Picenum hardly explain the alias. It is likely to contain some derogatory allusion (cf. 27 (II. 7). 3 n. (*Pisaurensi*)), unless it is only a glossator's attempt to explain *fratribus*, based on *Firmano* above. Or the brothers Pulchri, Appius and Publius, may be meant: cf. 75 (IV. 3). 4 *fratrum trium* (*trium* includes Metellus Nepos, now in Spain); 86 (IV. 11). 2 *quid Appius, quid illa populi Appuleia.*

quod eram exorsus This can hardly refer to a previous letter: cf. 299 (XIII. 28). 2 *sed ut coepi*; *Acad.* II. 129 *sed quod coeperam.* C. seems to have slipped, as Quintilian apparently did when he says *ut dixi* of a remark which in fact does not occur in what precedes (*Inst.* VI. 3. 9). Caesar's *ad propositum revertar* (172 A (IX. 6 A)) is rather similar. Or just possibly Luscus and the brothers were connected in C.'s mind with the law mentioned in §1. Anyhow I do not feel forced to resort to Schütz's ingenious but technically improbable interchange of *quod quaeris* and *quod eram exorsus*. Sure cases of such shuffling in the MSS are few and far between.

6 **Firmano** There is epigraphical evidence for Gavii at Firmum under the Principate, also in Tuscany and Verona: cf. Constans, *l.c.* *hoc* is not 'celui d'ici', at least not necessarily: cf. 23 (II. 3). 3 *hunc dico Balbum* et sim. Obviously the Gavius of 117 (VI. 3). 6 is a different man.

9 **removerit** C. nearly always has the subjunctive in such cases (cf. K.–S. II, pp. 270 f., 276), though *Verr.* II. 4. 43 *quid erat quod confirmabat* (just after *quid erat quod...quereretur*) seems to be an exception.

4, 1 **πολιτικῶς** 'Like a politician', almost 'diplomatically'; not, in this context, 'like a statesman' or 'good citizen'.

1–2 **τὴν ἔσω γραμμήν** A reference to the game of πεττεία, in which the two players each disposed their pieces on five rows: cf. A. S. F. Gow on Theocr. VI. 18. The rules are unknown, but the 'inner line' seems to be identical with the μέση or ἱερὰ γραμμή between the two 'territories'; this was apparently a place of safety on which stood the king (βασιλεύς), and defeated players (or pieces?) were sent to the end of it: cf. Eustath. 633. 59 ὁ ἡττώμενος ἐπ' ἐσχάτην αὐτὴν ἵετο. Hence the proverb τὸν ἀφ' ἱερᾶς κινεῖν for bringing up one's last resources. The gist of Atticus' advice was therefore that C. should not take risks; he should stay safe, like a πεττός on the ἱερά. I strongly suspect that the key lies hereabouts to a puzzling expression in Ter. *Eun.* 640 *certe extrema linea / amare haud nil est* (cf. Otto, *Sprichwörter*, p. 194).

2 **prudentia** Often of knowledge or skill in a particular *métier*: cf. *de Orat.* I. 256 *prudentiam iuris publici*; *Phil.* II. 81; Liv. I. 54. 1, etc. In this context it almost amounts to 'information'.

4 **convivam** Who this was we cannot tell; probably someone like Theophanes, in close touch with the powers that were (cf. 30 (II. 12). 2).

degustes Cf. *ibid.* ὀξύπεινος. Seneca (*Ep.* 11. 1) has *dedit nobis gustum*, 'gave us a taste of his quality'.

83 (IV. 6)

1, 1 **Lentulo** L. Lentulus Niger: cf. 44 (II. 24). 2 n.

1–2 **magnum hominem** Not 'fine fellow' or 'homme de cœur': cf. *Rep.* II. 12 *consilium...magni hominis et iam tum longe providentis*; *Acad.* II. 117 *est enim inter magnos homines summa dissensio* et sim.

5 **Saufeius** Cf. 8 (I. 3). 1 n.

vestri I.e. *Epicurei*.

8 **natura πολιτικός** Cf. *Fam.* II. 8. 1 (to Caelius) πολιτικώτερον *enim te adhuc neminem cognovi*. The contrast between Atticus, who kept aloof from public life by his own choice, and C. himself is hinted with characteristic delicacy.

9 **†fueris non ne** Wesenberg's *servis nomine* ('communem habes servitutem') may be right. But *communi nomine* thus is not exactly paralleled, though *suo nomine* = 'on one's own account' in certain contexts: cf. *Verr.* II. 4. 72 *cum illa civitas cum Poenis suo nomine ac sua sponte bellaret*; Liv. XXXIV. 31. 5 *cui et publice...vobiscum...foedus sit et meo nomine privatim amicitia ac societas.* For sense *uteris condicione* (Koch: cf. *Cael.* 57 *communi condicione servitutis uti*) is satisfactory. The irony of *frueris nomine* (Pius) would be inept and offensive.

2, 5 **ingratus** For his restoration.

6 **nequiquam** C. foresaw that the dynasts would want more from him than mere neutrality.

6–7 **immo...in castra** Sc. *ibimus. immo etiam* (cf. *Thes.* V (ii). 953. 31) would not introduce a question.

9 **reliquum est** Orelli's superfluous *iam* does not really help palaeographically. There are other possibilities of course, e.g. *quid reliqui est?* (cf. Ter. *Heaut.* 193), or *reliqui nil est* (cf. *Fin.* I. 30 *reliqui nihil est*).

Σπάρταν 'Sparta' is one's allotted portion or task, to be made the best of; in 20 (I. 20). 3 the politics of the *via optimas*, here those of collaboration.

10 **Philoxeno** According to the story (in Diod. XV. 6. 2 f.; cf. Suidas (Adler) I, p. 258), the poet Philoxenus of Cythera was sent to the quarries by Dionysius I of Syracuse for criticizing his tragedies. On being recalled and asked to praise them he simply told Dionysius' men to take him back to the quarries. ἄπαγέ με εἰς τὰς λατομίας became a proverb (cf. Leutsch–Schneidewin, *Corp. Paroem. Graec.* I, p. 398).

12 **ista probem** The whole context demands that *ista* be understood of the doings of the dynasts, like *illa causa* in 39 (II. 19). 2 *ego autem neque pugno cum illa causa propter illam amicitiam neque approbo ne omnia improbem quae antea gessi. probem* (or *approbem*, or *iam probem?*) is better than ⟨*ne*⟩ *improbem* (Schütz)—active collaboration is the pill to be swallowed, not mere abstention from disapproval. With *istam* (sc. Σπάρταν, Lambinus) we should rather expect *ornem.*

17 **ferrei** A fully acceptable correction, on which modern efforts (*servi, miseri, miserrimi*) do not improve. *ferrei* provides the most forcible of contrasts to *non miser.* The living are not merely unhappy, but so unhappy that it is a wonder they can bear to live at all—'nimis patientes, ἀπαθεῖς, qui adhuc vivamus nec voluntaria morte nos (a servitute) liberemus' (Orelli). That is how the words must be taken: not 'but we who are so callous as to live on' (Shuckburgh, supplying *miseri sumus*); much less 'c'est nous qui le sommes, qui avons une santé de fer' (Constans). Kayser's notion that *ferrei* alludes to Hesiod's Iron Age belongs in limbo. *ferreus*, a favourite word with C., is a *mot juste* in such a context, as Shuckburgh went some way to

show by citing *Amic.* 87 *quis tam esset ferreus qui eam vitam ferre posset?* (the jingle *ferreus–ferre* is probably fortuitous). For an even closer parallel go to Ov. *Met.* XIII. 516 where Hecuba asks over Polyxena's dead body *quo ferrea resto?* Similarly *crudelis* (*ibid.* XI. 701: *crudelior ipso / sit mihi mens pelago, si vitam ducere nitar / longius*), *durus* (Tib. (Lygdamus) III. 2. 3 *durus et ille fuit, qui tantum ferre dolorem, / vivere et erepta coniuge qui potuit*), *sceleratus* (*Cons. Liv.* 135 *tene ego sustineo positum scelerata videre?*). Had C. written *nos vero ferrei qui vivamus* I imagine few would have complained. The question is then whether *qui vivamus* can fairly be left to the reader's imagination. Surely it can. C. has just said that Lentulus is lucky to be taken: *nam quid foedius nostra vita, praecipue mea?* After that the contrast between Lentulus, lucky to be dead, and *nos*, the living, is enough. Atticus would need no plainer pointer. Add that the misery and dishonour of surviving the Republic is a commonplace in these letters: cf. 147 (VII. 23). 2 *manebo igitur, etsi vivere*—; 179 (IX. 12). 3 *populi Romani exercitus Cn. Pompeium circumsedet...fuga prohibet: nos vivimus*; 299 (XIII. 28). 2 *quam enim turpis est adsentatio, cum vivere ipsum turpe sit nobis*; 363 (XIV. 9). 2 *itaque quam severe nos M.* †*Cutius*† *accusat, ut pudeat vivere, neque iniuria*: also Quint. *Inst.* VI. praef. 14 *nemo nisi sua culpa diu dolet. sed vivimus.*

3, 1 **Hortensiana** Little is known about C.'s relations with Hortensius since his return, except that they appear to have co-operated in the Senate over the matter of Ptolemy's restoration (*Fam.* I. 5*b*. 2, 7. 2) early in 56, and in March jointly defended P. Sestius. This suggests that there had been no open breach up till then (cf. C.'s polite reference in *Sest.* 3) despite C.'s bitter resentment of Hortensius' supposed treachery in 58. From the present passage it is clear that they had since quarrelled; *illius amici* implies that Hortensius was *amicus* no longer. The *iniuria* was not, as usually supposed, Hortensius' conduct in 58, but something more recent; for, apart from the fact that this was now rather ancient history, *intemperiem* would be entirely inappropriate to the perfidy masquerading as friendship of which Hortensius had then (in C.'s eyes) been guilty. The later breach could easily have arisen out of Hortensius' support for Clodius (cf. 75 (IV. 3). 3 n. (*familiari tuo*); *Fam.* I. 9. 19). Atticus, as a common friend, naturally sought peace, and to that end urged C. to publish something emollient (the notion that the proposed work was to be an attack on Hortensius is plainly absurd). Miss L. R. Taylor (*Class. Phil.* 44 (1949), pp. 219 f.) sees in this the genesis of the *de Oratore*, actually finished in November 55 (cf. 87 (IV. 13). 2).

alia Possibly, as Miss Taylor suggests, *commentarii* on the events of 63 for Lucceius to work up: cf. §4 and *Fam.* V. 12. 10.

2 **incipiendo** Almost equivalent to *incipiens*: cf. K.–S. I, pp. 752 ff. (add 200 A (X. 9 A). 4 (Caelius) *quod tu non dicendo mihi significasti*). If there

is any difference, it seems to lie in the closer connection implied between the two verbs—the attempt and the withdrawal were so nearly simultaneous as to be virtually one action. *in* or *ab* should not be added.

3 **tulisse** The omission of *non* suits C.'s self-congratulation on the score of βαθύτης and gives a better balance with *iniuriam...illustrem* than the vulgate *intemperiem non tulisse* (Malaespina). For *non* added before *in* cf. 43 (II. 23). 3 *ingrederis*; 126 (VII. 3). 5 *invidere*; also Müller, p. xxvi. *modo* (*m̃*) is another possibility. *illius amici* alludes to the period before the quarrel; people, says C., had thought him too long-suffering.

5 **βαθύτης** 'Die Fähigkeit, das, was wir etwa Animosität nennen, aber auch tiefer liegende Entrüstung und Erbitterung zurückzudrängen, nennt Cicero im Briefstil βαθύτης' (F. Zucker, *Philol.* 93 (1938), p. 36): cf. 103 (V. 10). 3; 115 (VI. 1). 2.

6 **occultior** *oculatior* (Bosius) reverses C.'s meaning; and the sense 'more conspicuous' has only the warrant of a second-century *senatus consultum* (Plin. *N.H.* XXXIV. 24 *quam oculatissimo loco*).

satisfactio Probably 'apologia': cf. Sall. *Cat.* 35. 2 *satisfactionem ex nulla conscientia de culpa praeponere decrevi, quam medius fidius veram licet cognoscas.* This suits *ne* βαθύτης...*occultior*. C. and Hortensius were certainly friends again by 51 (95 (V. 2). 1).

levitatis There would be more *gravitas* in simply holding one's peace and going one's way than in such literary gestures.

4, 3 **epistulam** *Fam.* V. 12.

quam Many texts have *nunc quam*, but *nunc* seems pointless. For *quam* / *numquam* in the MSS cf. 16 (I. 16). 11 *quam reliquisti.* Mr C. P. Jones in my seminar at Harvard suggested *nostro* (*nr̄o*).

meas res Sc. *gestas.* So *Fam.* V. 12. 1 *res nostras.* Lucceius had already almost finished a history of the Social and Civil Wars and proposed to continue it (*Fam.* V. 12. 2). But he seems never to have published this, nor to have kept his promise to C.

6 **domum** Cf. 84 (IV. 10). 2; 85 (IV. 9). 2.

Vestorio C. Vestorius of Puteoli, much concerned in C.'s business affairs and, at least later, something of a personal friend of his as well as of Atticus' (cf. *Fam.* VI. 11. 2 *noster familiaris*).

84 (IV. 10)

1, 1 **Ptolemaeum** Ptolemy Auletes had lost his throne in 58. The question of whether, how, and by whom he should be restored loomed large in 57–56. It had just been settled by Gabinius, governor of Syria, at Pompey's instigation (cf. Dio, XXXIX. 55. 2, 56. 3).

2 **pascor** Cf. 86 (IV. 11). 2 *nos hic voramus litteras.*

3 **Fausti** Faustus Cornelius L. f. Sulla. Distinguished service under Pompey (whose daughter he married) as military Tribune (63), a great celebration of games in honour of his father the Dictator (60), and election to the Augural College (before 57) had so far marked his career. Quaestor in 54 and an active republican in the Civil War, he was captured and killed along with Afranius by Caesar's men after Thapsus. Being in financial difficulties (cf. 178 (IX. 11). 4; Plut. *Cic.* 27), he appears to have sold his library to C. It will have mainly consisted of books which his father brought back from Greece and Asia (cf. Plut. *Sull.* 26).

his rebus It is unlikely that *res* means 'fare' specifically in Hor. *Sat.* II. 6. 111 or any of the other passages quoted by commentators. But *pascor* shows what C. had in mind, the famous products of the Bay (cf. Varro, *Menipp.* (Bücheler) 160 *patella esurienti posita provocat Neapolitanas piscinas*; Plin. *N.H.* IX. 169, etc.). To give *pascor* a metaphorical sense with both substantives (cf. *Pis.* 45 *his ego rebus pascor, his delector*) dissolves the joke. *ostreis* (Ursinus) is a better scholion than emendation.

5 **et voluptatibus sqq.** Simple omission of *et* leaves *voluptatum* as gen. epexegeticus after *oblectationibus*, a totally unconvincing result. Lehmann proposed *prope taedet* before *propter*, but a reference to C.'s age is a desideratum. He must not be left to imply that *only* political griefs took away his appetite for oysters.

8 **istorum** Pompey and Crassus.

tecumque...ambulare Cf. *Fam.* II. 12. 2 (to Caelius) *cum una mehercule ambulatiuncula atque uno sermone nostro omnis fructus provinciae non confero.*

8–9 **cum eo** With Pompey (cf. §2). *ambulandum* is *not* metaphorical.

10 **qui curet** Perhaps with a glance at Atticus' Epicureanism: cf. *Div.* II. 104 *Epicurusne? qui negat quicquam deos nec alieni curare nec sui*; Quint. *Inst.* V. 6. 3 *philosophi...qui deos agere rerum humanarum curam negarent.* Ribbeck found a senarius in *fors...deus* (reading *qui est*).

2, 2 **Laconicum** A kind of sweating bath. The information that such baths were first used by the Lacedaemonians comes from Lewis and Short, not Vitruvius: cf. K. Schneider, *RE.* XII. 347. 68. The word first occurs in a Pompeian inscription of the Sullan period (*C.I.L.* I. 1635, X. 829).

circa The vulgate *Cyrea* (Victorius) *sint* ('talia quae sint e genere Cyreo', Müller) has been too long with us. Why should buildings in the style of Cyrus (cf. 23 (II. 3). 2 n.), supposing he had a style, need Atticus' attention more than others? On *sunt* in E (also conjectured by Wesenberg) see A. C. Clark, *Cl. Rev.* 33 (1919), p. 40.

quoad poteris Cf. 1 (I. 5). 7 n.; 83 (IV. 6). 4 *domum nostram quoad poteris invisas.* The personal form seems much better suited to such a request.

3 **Philotimum** Cf. 24 (II. 4). 7 n.

4 **respondere** Cf. 75 (IV. 3). 6 n. For *aliquid*, 415 (XVI. 7). 6 *ut sit unde par pari respondeam*.

5 **Parilibus** 21 April.

85 (IV. 9)

1, 1 censum The election of Censors had been expected in 56 (cf. 74 (IV. 2). 6), five years after the last elections in 61 (cf. 17 (I. 17). 9 n.), but the disorders of that year made it impossible. Pompey and Crassus, who after their own election in January 55 held *comitia* for the other magistracies (cf. Broughton, p. 214), held the censorial elections in which Messalla Niger and Servilius Isauricus were successful. They were still in office, with some time to run, in July 54, but did not complete the lustrum (cf. 89 (IV. 16). 8). Why the Tribunes should have wished to obstruct, or what *totaque de censura* connotes, remains unknown. The trouble may have been connected with Crassus' levying of troops: see on 87 (IV. 13). 2.

diebus vitiandis By *obnuntiatio*: cf. 75 (IV. 3). 3 n. (*de caelo servaturum*).

3 **hic** I.e. *in his locis*.

multum Constans compares 163 (VIII. 13). 2 *multum mecum municipales loquuntur*.

5 **Syriam** Assigned to Crassus, as Spain to Pompey, by the lex Trebonia, which evidently had not yet been passed—Pompey would not be scorning a province which had already been assigned to his colleague.

iactans I.e. *fastidiens*: cf. 227 (XI. 16). 3 *sin iactor* and *Thes*. VII. 55. 70. 'Boasting' is improbable after *sibi displicens*.

7 **καὶ τόδε Φωκυλίδου** As prefixed by Phocylides to his γνῶμαι: cf. Diehl, *Anth. Lyr. Gr.*[3] I, fasc. 1, p. 57.

10 **a. d. VI** = *ad sex*. C. would naturally give the date. Earlier conjectures (*a se*, *etsi*, etc.) can be discarded.

10 **Messallam** Cf. 23 (II. 3). 1 n.

2, 1 commendaturum As a theme: cf. 83 (IV. 6). 4.

5 **Paetum** Cf. 20 (I. 20). 7 n.

86 (IV. 11)

1, 3 Demetrio Of Gadara, Pompey's favourite freedman; presumably therefore Cn. Pompeius Demetrius. Various stories are told of his wealth and hauteur. Probably he was the bore whose company C. was anxious to avoid ten years later: cf. *Fam*. XVI. 17. 2, XVI. 19 (see *Proc. Camb. Phil. Soc.* N.S. 5 (1958–9), pp. 11 ff.).

5 **esse statim venturos** Cf. 151 (VIII. 1). 2 *statim esse venturum*. *et se* ⟨*et eum*⟩ is equally possible. Editors read ⟨*eum*⟩ *et se* (Lehmann), but the speaker normally has precedence.

5–6 **ut...putarent** Possibly this refers (ironically?) to a project of provincial tax-reform: cf. F. E. Adcock, *Camb. Anc. Hist.* IX, p. 615.

2, 2 **Dionysio** Cf. 82 (IV. 8*a*). 1 n.

οὐδὲν κ.τ.λ. From Menander, play uncertain; perhaps *Epitrepontes* (fr. 2, Körte): cf. Kock, *Com. Att. Frag.* III, p. 227. The line is quoted elsewhere with slight variations, which may point to οὐδὲν γλυκύτερόν ἐστιν (Cobet) in the first half.

4 **quid primus dies** Of the gladiator show. The same verb cannot be supplied with *dies* as with *censores*, etc., so probably *quid* amounts to 'all about': cf. 30 (II. 12). 4 n. (*quid Caesar?*).

Appius Now standing for the Consulship. He was elected.

5 **illa populi Appuleia** Clodius, the modern Appuleius Saturninus. *Appuleia* contains an insult like *filiola Curionis* in 14 (I. 14). 5: cf. Orelli on Hor. *Sat.* I. 8. 39 *Pediatia*.

8–9 **ab isto tantopere** Cf. *Arch.* 12 *quaeres a nobis, Gratti, cur tanto opere hoc homine delectemur* (also 416 (XV. 13). 3; *Verr.* I. 35; *Pis.* 59). Haplography produced *abistopere*. Constans reads *ab isto magno opere* (Tyrrell), influenced perhaps by *abs te quo magnopere* in ancient editions. Other conjectures include *Aristotele magno opere* (Orelli), *ita ab isto puero* (Madvig), *ab isto opipare* (Ziehen), ἀποθεωρήσει (Gurlitt).

9 **librum** Perhaps notes for a history of C.'s Consulship: cf. 83 (IV. 6). 3 n. (*alia*).

10 **Demetri Magnetis** A contemporary littérateur, whose book περὶ ὁμονοίας is mentioned in 161 (VIII. 11). 7. There is no particular reason to think that this is the same work.

statim Atticus will have asked C. to send the book along at his convenience. Editors generally take with *referat*, putting a comma after *mitto*.

87 (IV. 13)

1, 2 **XIII** Or *XIIII*.

3 **quod dico** Cf. *Off.* III. 110 *nam quod aiunt quod valde utile sit, id fieri honestum, immo vero 'esse' non 'fieri'*. The vulgate *quid dico 'volumus'?* may of course be right: cf. *Thes.* VII. 475. 67.

4 **nuptiae** Milo was about to wed Fausta, twin sister of Faustus Sulla and divorced wife of C. Memmius (cf. Ascon. 28. 20).

comitiorum Evidently the elections of 55 had been postponed, like those of the previous and the two following years, but the circumstances are unknown.

ego sqq. Many texts have Bosius' *ego, ut sit rata, afuisse* sqq., as though the *prospect* of elections would make it any the less reasonable for C. to congratulate himself on having missed the unpleasantness in the Senate. Nor are subsequent guesses any happier. The lacuna here postulated may have contained something like *desiderari me nuntiant.* For *pro re nata* cf. 131 (VII. 8). 2; 138 (VII. 14). 3; 360 (XIV. 6). 1.

5 **altercationibus** We have no direct evidence as to the subject of these, but probably they concerned disputes over Crassus' *ornatio provinciae*: cf. Dio, XXXIX. 39. In that case *cui non oporteret* in the next sentence probably means Crassus, with whom C. was now officially on good terms (cf. *Fam.* V. 8, I. 9. 20), not Pompey.

9 **σκυλμόν** 'Annoyance'; in medicine 'irritation'.

10 **suspecta** εὔπεπτα (W. B. Sedgwick) is at least a most ingenious suggestion.

2, 2 **paludatum** Wearing the *imperator's* cloak (*paludamentum*), the term regularly applied to a Consul leaving Rome for a military command: cf. Mommsen, *St.* I, pp. 431 f.

3 **L. Paulum** L. Aemilius Paulus, Cos. 182 and 168, left for Macedonia in the latter year *maiore quam solita frequentia prosequentium* (Liv. XLIV. 22. 17). Crassus had made himself unpopular by levying troops, and his plans for war against Parthia were bitterly attacked by opposition Tribunes, one of whom, Ateius Capito, solemnly cursed him (cf. Dio, XXXIX. 39. 5; Plut. *Crass.* 16). He was now 59 or 60 years old.

5 **libris oratoriis** Cf. 83 (IV. 6). 3 n. (*Hortensiana*). The *de Oratore* was now completed. So too C. says a few weeks later in his letter to Lentulus Spinther (*Fam.* I. 9. 23).

6 **etiam** *etiam ⟨atque etiam⟩* Schütz, perhaps rightly: cf. 365 (XIV. 11). 2; 162 B (VIII. 12 B). 2 (Pompey); *Thes.* V (ii). 927–9.

7 **τυπωδῶς** Sc. *scribe.*

hospes Cf. 117 (VI. 3). 4 *est enim quiddam advenientem non esse peregrinum atque hospitem*; *Fam.* II. 12. 1 *ne hospes plane veniam*; *Mil.* 33 *vos hospites in hac urbe versamini?* et sim.

88 (IV. 14)

C. passed all or much of May 54 in his Campanian villas: cf. *Q.Fr.* II. 13. 1. Shortly before he wrote, Atticus had left Rome for Epirus. As T. Schiche saw (*Jahresber. d. Phil. Ver.* 34 (1908), p. 59), C.'s request in §2 for news from his brother and Caesar shows that Atticus was travelling by way of North Italy, where Quintus, who had left Rome about the end of April to join Caesar's staff (cf. *Q.Fr.* II. 13. 1), and apparently Caesar himself were to be found (cf. *ibid.* p. 13; *Q.Fr.* II. 14. 1).

1, 1 **Vestorius** Cf. 83 (IV. 6). 4 n.

2 **putare** Sc. *se.* The text has been much solicited; but if not very elegantly expressed, C.'s meaning seems not too hard to establish. The reason for the delay (not the departure itself) was only Vestorius' opinion—he had not reported that as a fact. Schiche proposed *Patavium.*

2 **dixeras** Sc. *mihi*. Obviously more likely than *dixerat*, though the latter is not untranslatable. Such confusions of personal endings are common.

3 **si...vales** As indicated by the fact that he had now set out.

4 **domum ad te** *ad te* makes it clear that 'home' means Atticus' home, not the writer's: cf. 249 (XII. 11) *domum ad se*.

5 **Varronis** Perhaps purchased from him: cf. 84 (IV. 10). 1 *bibliotheca Fausti*.

6 **eos quos...habeo** The books *de Republica* (cf. 89 (IV. 16). 2). For the changes of plan contemplated and in turn rejected see Philippson, *RE*. VII A. 1109 f.

2, 2 **a C. Caesare** This need not imply that Atticus expected to meet Caesar or even to hear from him *directly*; *a* can be simply 'from the quarter of', as probably in 200 (X. 9). 3 *si quid ab Afranio*. It is doubtful whether Atticus had any personal relations with Caesar at this time that went beyond mere acquaintance, and even this is not definitely attested: see Intr. p. 7, n. 1.

6 **feliciter ex sententiaque** Cf. *Fam*. II. 7. 3 *quoniam sane feliciter et ex mea sententia rem publicam gessimus. totoque itinere ex sententia confecto* of the old editions is well enough *per se* but would have offered the copyist no such invitation to a slip as *itinere* followed by *-icitere*(*x*).

89 (IV. 16)

The letter was written between 28 June, the date of Scaurus' arrival in Rome (Ascon. 18. 18; cf. §6), and 3 July (§5); Atticus will have been in Epirus for some time, having set out on 10 May (88 (IV. 14). 1).

1, 2 **librari** Cf. 40 (II. 20). 5 n. and *Q.Fr*. III. 3. 1 *occupationum mearum tibi signum sit librari manus*.

3–4 **quod erant abs te** 'Tam inanes erant ut nihil fere aliud continerent nisi simplicem loci unde scriptae essent significationem, vel si quid amplius, illud unum, te recte valere' (Orelli). The perplexities which have led some editors to follow Schütz in placing these words after *recte esse* (where they become nonsense) should disappear once it is recognized that *abs te* cannot mean *tua manu* and does mean 'from your residence', i.e. from Buthrotum (or possibly also from *deversoria en route*: cf. 422 (XVI. 10). 1 *veni ad me in Sinuessanum*). So one might say, 'Your letters tell me *where* you are, by the home address, and sometimes also *how* you are'. Atticus probably moved on from Epirus fairly soon (cf. 90 (IV. 15). 3).

5 **abs te Buthroto** *Buthroto* is added for clarity; it suggests, though it does not perhaps necessarily imply, that some of the previous letters had been addressed from other residences. Cf. 422 (XVI. 10). 1 (see above); 88 (IV. 14). 1 n. *domum ad te*.

8 **gravis** Cf. 13 (I. 13). 1 n. *graviorem*.

8 **M. Paccius** There were Paccii at Formiae and Minturnae (cf. Münzer, *RE*. XVIII. 2064. 42). Paccius had presumably been Atticus' guest (not host) in Epirus on his way back to Italy.

2, 1–2 **Varro...si modo erit locus** Historical fiction in the *de Oratore* and the *de Republica* excluded living persons, except for the proems: cf. Q.*Fr*. III. 5. 1; 324 (XIII. 17–18). 2; 326 (XIII. 19). 3 f. Varro is not mentioned in the surviving portions of the *de Republica*, and discussion of his role in the *Academica* ten years later (326 (XIII. 19). 3) suggests that no *locus* was found for him in the political dialogue. Nor did C. finally change the plan of the *Academica* in accordance with the letter to Quintus, as would have enabled him to introduce contemporary speakers.

6 **in Africani personam** The *dialogi personae* here as in *Q.Fr.* III. 5. 1 are those of the published work, with the single omission of L. Mummius (on which see Philippson, *RE*. VII A. 1109. 56).

10 **ἐξωτερικούς** Aristotle's published works, ἐξωτερικοὶ λόγοι, chiefly the dialogues as distinct from the treatises: cf. H. Bonitz, *Index Aristotelicus*, pp. 104*b* f.; W. Jaeger, *Aristotle*, pp. 31 ff., 246 ff.; P. Moreaux, *Les Listes anciennes des ouvrages d'Aristote*, pp. 167 ff., 300 f., 307.

3, 1 **Scaevolae** Q. Mucius Scaevola the Augur, Cos. 117 (the Scaevola of the *de Republica* is the same man). He was nearly if not quite eighty at the supposed time of the dialogue *de Oratore* (91 B.C.), and died some four years later. C. saw much of him (*Amic.* 1).

2 **idem quod** 'In archetypo videtur fuisse *sed feci quod in* super *quod* adscripto *idem*' (Sjögren).

deus Cf. A. S. Pease on *Nat. Deor.* II. 32 *Platonem, quasi quendam deum philosophorum.*

3 **in Piraeum** On the preposition cf. 126 (VII. 3). 10 n.

5 **adest in disputando** It is rather curious that C. should think it needful to remind Atticus in such detail of one of the most familiar scenes in Plato; nor does he do it quite accurately. The *Republic* virtually begins with the duologue on old age between Socrates and Cephalus, after which Cephalus withdraws.

9 **magis** *satis* in the MSS is obviously a reflection of *satis consonum* above or *satis decorum* below.

10 **qua esse** *qua eum esse* Bücheler, perhaps rightly in view of the paradosis *quam*.

11 **decorum** A sidelight on Roman ideas of dignity. C.'s feeling in 44 that the 'labour of a sea voyage' (413 (XVI. 3). 4) was inconsistent with his position (*dignitas*) as well as his age has something of the same quality.

14 **ioculatorem** There is no good reason to suspect this word, which, like a number of others in these letters, is unique in classical writing. To

Roman sentiment there might well seem something incongruous, even ridiculous, in early-first-century grandees expounding the technicalities of the rhetorical schools, so that Scaevola, if present, would be apt to make fun of the proceedings (so C. suggests, as though the dialogue was a real conversation arranged by himself). Scaevola's jocularity comes out in his Parthian shot at Antonius, *de Orat.* I. 265 *simul adridens* sqq.

4, 1 **re Piliae** Its nature, like Aurelianus, is unknown. *luculenta* suggests a piece of property or an inheritance (cf. *Phil.* XII. 19 *luculenti patrimoni*).

2–3 **me...venditabo** Either because of Tullia's partiality for Pilia or for some other reason arising from the nature of the *res.*

4 **sed scis qui** Or perhaps, as some punctuate, *sed (scis qui) cum* sqq.: cf. Plin. *Ep.* IV. 15. 1 *Cornelium Tacitum (scis quem virum).* But it may be that *sit* (Wesenberg) should be added: cf. Ter. *Andr.* 919 *videat qui siet.* Or did C. write *sed scin quid?* (cf. Plaut. *Poen.* 1167, *Pseud.* 641 *scin quid est?*, *Poen.* 376 *at scin quo modo?* et sim.), or *sed scis tu* (cf. Id. *Bacch.* 202)?

5, 1 **C. Catone** C. Porcius Cato probably belonged to the elder branch of the family, descended from the Censor's first wife Licinia. In 59 *adulescens nullius consili, sed tamen civis Romanus et Cato* (*Q.Fr.* I. 2. 15), he tried to prosecute Gabinius on a bribery charge and barely escaped alive after speaking against Pompey at a public meeting; but as Tribune in 57–56 he emerges as a backer of Clodius and protégé of Crassus in conflict with both Milo and Pompey, and particularly conspicuous in opposition to Lentulus Spinther's claims (which C. supported) to restore King Ptolemy (*ibid.* II. 1. 2, 3. 3 f., 5. 3 f.). Reconciled with Pompey later in 56, he held up the elections in his and Crassus' interest (Liv. *Epit.* CV; Dio, XXXIX. 27. 3). His turbulence earned him these prosecutions; as C. predicted he was acquitted in the second (cf. 90 (IV. 15). 4). What became of him thereafter is unknown.

1–2 **Iunia et Licinia** Cf. 29 (II. 9). 1 n.

2 **Fufia** Cf. 16 (I. 16). 13 n. W. F. McDonald (*Journ. Rom. Stud.* 19 (1929), p. 179, n. 2) points out that there was also a lex Fufia iudiciaria, passed by Fufius Calenus in 59: cf. Schol. Bob. (Stangl), p. 97; Dio, XXXVIII. 8. 1. But Cato's activities as Tribune, which were clearly the basis for the prosecution (cf. 90 (IV. 15). 4), are not likely to have come within its scope.

3 **neque...accusatoribus** I.e. there was *praevaricatio*, perhaps through Pompey's influence.

4 **tamen** I.e. 'I think he ought to be convicted, but...'.

Drusus Cf. 27 (II. 7). 3 n. He was accused of *praevaricatio* (*Q.Fr.* II. 16. 3; cf. 91 (IV. 17). 5) and acquitted, Cicero defending, about two months later; a speech against him by Licinius Calvus is mentioned in Tacitus' *Dialogus* (21. 2). The whole passage down to *iudicia nosti* hangs together if we suppose that Drusus was one of C. Cato's prosecutors, and was himself prosecuted for collusion in that capacity (which suits his previous political record) by Lucretius and Calvus. Both were 'anti-triumviral' men, and the latter himself accused Cato in one of his two trials (cf. 90 (IV. 15). 4 n.). Scaurus, a supporter of Pompey's, defended him in the second (under the lex Fufia), which ended with another acquittal on July 4 (Ascon. 18. 16).

5 **Lucretio** Probably Q. Lucretius, the Senator and friend of C. Cassius who surrendered Sulmo to Antony in 49 (cf. 148 (VII. 24); 156 (VIII. 4). 3; Caes. *B.C.* I. 18. 1). He in turn is identified by Miltner (*RE.* XIII. 1691. 47) with the Q. Lucretius Q. f. Vespillo who commanded a Pompeian squadron in the Adriatic (Caes. *B.C.* III. 7. 1) and escaped the proscriptions to receive a Consulship from Augustus in 19 B.C. (on the vexed question of the *laudatio Turiae* see Münzer, *RE.* XIII. 1389). The cognomen Vespillo derived from a grandfather (?), who earned it as Plebeian Aedile in 133 by ordering Ti. Gracchus' body to be thrown into the Tiber (Val. Max. I. 4. 2; *Auct. Vir. Ill.* 64. 8).

6 **Procilio** Also prosecuted, no doubt in connection with the disturbances in 56: cf. 90 (IV. 15). 4 n. *non boni* is obviously ironical; in C.'s opinion Procilius ought to be convicted, but juries being what they were, he might escape.

Hirrus C. Lucilius C. f. Hirrus, of senatorial but not noble family (cf. Caelius, *Fam.* VIII. 2. 2 *nobilem agentem*), grand-nephew (probably) of the satirist, cousin and faithful follower of Pompey, great landowner and fish-fancier (Varro, *R.R.* II. 1. 2, III. 17. 3), and in C.'s opinion a conceited ass (cf. *Q.Fr.* III. 6. 4 *o di, quam ineptus, quam se ipse amans sine rivali!*). Now a candidate for the Tribunate which he held in 53, he earned C.'s animosity by standing against him for the Augurate in that year. Hence much glee over his failure in the aedilician elections of 51 (cf. *Fam.* II. 9. 1, etc.). Pompey's envoy to Parthia in 48, he lived to lend Caesar 6000 lampreys for his triumphal banquet in 45 (Plin. *N.H.* IX. 171) and to flee proscription in 43–42. His *rapprochement* with the Consul Domitius, so bitter an opponent of Pompey in 56, seems a foreshadowing of things to come. After his election as Tribune Hirrus began to work for Pompey's appointment as Dictator (*Q.Fr.* III. 6. 4).

7 **hi consules** L. Domitius Ahenobarbus and Ap. Claudius Pulcher: cf. *Scaur.* 27 *apud hos consules*. The provisions of the senatorial decree are unknown.

6, 2 candidatos Two patricians, Messalla Rufus and Scaurus, and two plebeians, Domitius Calvinus and Memmius. The lex Licinia required the election of one plebeian, and C.'s language both here and in *Scaur.* 34 discountenances T. J. Cadoux' theory (cf. Broughton, *Suppl.* p. 19) that this provision was no longer binding in the first century. C. cultivated all four (cf. *Q.Fr.* III. 1. 16). After unparalleled scandal and delay Messalla and Calvinus entered office in July 53.

Scaurum M. Aemilius M. f. Scaurus, son of the *princeps senatus* and that Caecilia Metella who later married Sulla. He himself married Pompey's divorced wife, Mucia Tertia. After serving in the East as Quaestor and governor of Syria, he became Curule Aedile in 58 along with another Pompeian lieutenant, Plautius Hypsaeus, and celebrated games with historic extravagance (Plin. *N.H.* XXXVI. 113, etc.). Praetor in 56, he governed Sardinia in the following year, whence this prosecution *de repetundis*. Acquitted with C. defending (cf. 91 (IV. 17). 4 n.), but unsuccessful in his candidature because of Pompey's desertion (cf. *Q.Fr.* III. 6. 3), he had to face a second prosecution, *de ambitu* (in 52 according to App. *B.C.* II. 24: cf. 91 (IV. 17). 5; Quint. *Inst.* IV. 1. 69). Again defended by C., he was condemned and vanished into exile. Asconius (18. 13) judged him his father's son for greed and arrogance, without his father's capacities.

3 **Triarius** P. Valerius C. f. Triarius, son of a Praetor but probably not one of the patrician Valerii: *adulescens paratus ad dicendum et notae industriae* (Ascon. 18. 20). He and his mother Flaminia were familiar friends of M. Cato's half-sister Servilia, a disquieting circumstance for Scaurus as Cato was president of the court (*ibid.* 19. 19). C. was on friendly terms with the whole family, especially with Publius' brother Gaius: cf. 267 (XII. 28). 3.

reum fecit Asconius says that Scaurus returned to Rome on 28 June (18. 18) and that the *postulatio* was on 6 July (19. 5). He seems to refer to the second stage of the prosecution, sometimes called *delatio nominis* (Sternkopf, *Hermes*, 40 (1905), p. 16: cf. Leifer, *RE.* XXII. 883).

6 **Domitius ut valeat** Early editions omit the second *ut*; the less usual order is a shade more likely to be original.—Whereas the Domitii Ahenobarbi took rank among the great families of the plebeian nobility at the beginning of the second century and retained it, the Calvini went into eclipse early in the third. They reappear with M. Domitius Calvinus, Pr. 80 and Sulla's governor of Hither Spain, whose son Gnaeus had a remarkable career. A conspicuously optimate Tribune in 59 and Praetor with optimate support in 56, in his present candidature he presumably followed the same line, in competition with the protégés of the dynasts, Memmius and Scaurus. But by the end of the year he had shifted his ground (cf. *Q.Fr.* III. 4. 1), and in 49 was the only Consular, apart from those convicted on criminal charges, to come out openly and actively for Caesar. As such he was naturally a prominent commander, holding Caesar's favour to the end, despite the fiasco at Nicopolis. A new career opened for him under Caesar's heirs, with a second Consulship in 40 and a Triumph in 36. He was probably still alive sixteen years later (cf. Münzer, *RE.* V. 1423. 67). Of the four candidates C. had least to do with him, yet *Domitio ipsi multa iam feci quae voluit quaeque a me petivit* (*Q.Fr.* III. 1. 16).

7 **adiuvetur sqq.** I.e. *munere etsi non gratissimo adiuvetur tamen*: cf. Housman on Luc. I. 333. *non ⟨nihil⟩* (Wesenberg), etc., deserve no consideration.

Memmius Cf. 18 (I. 18). 3 n.

8 **gratia** A copyist who had just written *Caesaris* might easily write *Gallia* for *gratia* or *grā*. *Pompei Gallia* could conceivably be understood as 'the part of Gaul controlled (electorally) by Pompey': cf. 114 (V. 21). 8 *in hac mea Asia*; Pollio, *Fam.* X. 33. 4 *ex Gallia Lepidi.* No doubt Pompey *had* influence particularly in Transpadane Gaul (cf. 104 (V. 11). 2 n.), and Cisalpine Gaul was a considerable electoral factor (cf. 10 (I. 1). 2 *quoniam videtur in suffragiis multum posse Gallia*). But his influence in Rome itself and elsewhere would be more important.

9 **adventum Caesaris** To Italy. He was now in northern Gaul, preparing for the invasion of Britain.

10 **absoluto** If C. Cato's obstruction in 56 had not gone unpunished, it might have been harder to find a Tribune to emulate him in 54.

7, 1 **Paccianae** Cf. 337 (XIII. 45). 1 *Diocharinae* (sc. *litterae*).

1–2 **fratris...Caesaris** C. had had these letters about a month. He received them on the day of his return to Rome from Campania, 2 June (cf. *Q.Fr.* II. 14. 1).

4 **exspectatur** In Rome. Caesar embarked on his second invasion of Britain, after delays due to bad weather, at the end of July (contemporary calendar), but of course the expedition was known to be in preparation long before: cf. *Q.Fr.* II. 14. 2; *Fam.* VII. 6. 2, 7. 1. I see no difficulty in *exitus.* One can await the outcome of what is about to take place as well as of something actually in progress; and for all they knew in Rome Caesar might already have sailed: cf. Sternkopf, *Hermes*, 40 (1905), pp. 18 ff.

5 **muratos mirificis molibus** The alliteration, the rare and metaphorical *muratos* (next in Vegetius), and the bold description of the cliffs of Dover (cf. Caes. *B.G.* IV. 23. 3) as *mirificae moles* stamp this phrase as mock-poetic, if not actually quoted from a tragic senarius such as *aditus murati mirificis sunt molibus.*

8 **musicis** Cf. *Pis.* 83 *servis symphoniacis et aliis muneribus acceptis* et sim.; Abert, *RE.* IV A. 1169 f. The references to Phemius and his κέρας in 113 (V. 20). 9, etc. show Atticus' interest in such commodities.

8, 1 **basilicam** According to Plutarch (*Caes.* 29: cf. App. *B.C.* II. 26) the Basilica Pauli, which L. Aemilius Paulus (cf. 44 (II. 24). 2 n.) beautified (προσεκόσμησε) as Consul in 50 out of funds received from Caesar, had been built ἀντὶ τῆς Φουλβίας, i.e. to replace the old Basilica Fulvia (or Aemilia et Fulvia; cf. Varro, *L.L.* VI. 4), erected in the Censorship of his ancestor M. Lepidus (179). The first of the two buildings mentioned by C. is evidently this old basilica and the second its successor, the magnificent Basilica Pauli (ὀνομαστὸν ἀνάθημα (Plutarch), οἰκοδόμημα περικαλλές (Appian: cf. Hülsen, *RE. Suppl.* I. 16)), with its pillars of Phrygian marble (Plin. *N.H.* XXXVI. 102), which was completed and dedicated by Paulus' son in 34 (Dio, XLIX. 42. 2). This was not 'right in the Forum' in 54, but bounded it to the north. See Constans (III, p. 39, n. 1), who rightly rejects (as did T. Frank, *Am. Journ. Phil.* 44 (1923), pp. 355 f.) the common view that the second building was the Basilica Julia. The text seems sound.

1 **texerat** An epistolary tense; the old building must have been removed at some later date. C.'s words will hardly cover Frank's explanation: 'Paulus had some time before this undertaken to repair the ancestral basilica and had almost completed the work on a modest scale when he heard that Caesar intended to erect buildings in front and behind which would inevitably overshadow his structure. Hence he decided to tear down the basilica and let contracts for an entirely new building.' He may have let contracts for the new basilica as Curule Aedile in 55 (cf. Münzer, *Hermes*, 40 (1905), p. 98; E. Badian, *Journ. Rom. Stud.* 49 (1949), p. 83).

4 **itaque** Implies 'not to be outdone'.

Oppium C. Oppius, *eques Romanus* and Caesar's friend and agent. He is usually mentioned in conjunction with Cornelius Balbus. C. was always on good terms with Oppius (see the letter to him in 44, *Fam.* XI. 29).

dirumparis Cf. *Dom.* 99 *dirumpatur licet ista furia...audiet haec ex me.*

5 **monumentum illud** The Forum Iulii. *monumentum* ('memorial') need not imply a 'positive building': cf. *Mil.* 17 *quod is in monumentis maiorum suorum* (i.e. *in via Appia*) *sit interfectus*; Prop. IV. 6. 17 *Actia Iuleae pelagus monumenta carinae*; Val. Max. III. 2. 14 *tui...excessus, Cato, Vtica monumentum est.*

6 **atrium Libertatis** Cf. Serv. *ad Aen.* I. 726 *alii atria magnas aedes et capacissimas dictas tradunt, unde atria Licinia et atrium Libertatis.* The building is often mentioned (cf. *Thes.* II. 1102. 75), but the exact site is unknown: cf. Platner–Ashby, *Topographical Dictionary of Ancient Rome*, pp. 56 f.

7 **sescenties** According to Pliny (*N.H.* XXXVI. 103) and Suetonius (*Iul.* 26. 2) the purchase of the site cost 100 millions or more.

9 **nam** Not explanatory, as modern texts which punctuate with a semi-colon suggest, but 'occupatory'; the buildings on the Campus were a separate and lesser matter.

saepta The Saepta Julia were continued after Caesar's death by Lepidus and completed in 26 by Agrippa: cf. Dio, LIII. 23. 1.

11–12 **villa...publica** Built in 435, restored and enlarged in 194 (Liv. IV. 22. 7, XXXIV. 44. 5), just north of the Piazza del Gesù (Platner–Ashby, *op. cit.* p. 581). It served as a rendezvous for various public purposes, also to house foreign envoys.

12 **quid...proderit?** If Atticus is supposed to speak merely as a

member of the public, C.'s reply is cynically facetious: he and Oppius were not concerned about the comfort of the citizenry but to glorify Caesar. But it is hard to see why the Saepta should not be considered a public benefit. Better to take *mihi* as Atticus *qua* individual; apparently he was not much in the habit of visiting the Campus Martius. A sense of humour frees C.'s flippant answer from any tinge of *rusticitas*.

13 **quid id laboramus?** Cf. *Sull.* 88 *id laborat.* Did C. write *laboremus*?

14–15 **lege Clodia** In 58 Clodius legislated to deprive the Censors of their power to expel members of the Senate, except when acting jointly and on a charge brought in their presence (Ascon. 8. 24: cf. Mommsen, *St.* II, p. 386). His law was annulled in 52. We do not hear elsewhere of expulsions by these Censors (cf. 85 (IV. 9). 1 n.), but Lange's conjecture is rendered highly probable by the mention of the lustrum immediately before. *Plotia* (*Plocia*) of some early editions would denote the lex Plautia *de vi*; but trials *de vi* would often be of political interest.

15 **fiunt** For *fiant* Sjögren cites the subjunctives in *Fam.* I. 5*a.* 1 *hic quae agantur quaeque acta sint* (*aguntur*...*sunt* Wesenberg) *ea te cognosse arbitror*, *ibid.* (Matius) XI. 28. 2 *nota enim mihi sunt quae in me*...*contulerint*—neither quite on all fours.

9, 2 **C. Decimius** A *quaestorius* of this name was in charge of stores on the African islet of Cercina in 46. He fled on the approach of a Caesarian squadron under the historian Sallust (*Bell. Afr.* 34. 2).

5 **a nobis** Palaeographically a little better than *a tuis* (Boot). The modern vulgate *ab his tot tuis* (Sternkopf, comparing 90 (IV. 15). 2) seems less satisfactory; an adverb of time is wanted to balance *longe*. For *totiens* cf. Prop. IV. 3. 2 *cum totiens absis*; Ov. *Her.* 19. 70 *cur totiens a me, lente morator, abes?* Here as in Propertius *totiens* practically amounts to *tam diu*, but *can* be understood strictly as 'on so many occasions (when your presence is called for)'.

7–8 **hoc...tantum** Cf. 18 (I. 18). 8 n. (*ex iis quae scripsimus tanta*). It seems very improbable that Bosius invented these words, which he attributes to the Tornesianus *only* and which he himself rejected as a gloss.

11 **is** For sense cf. 90 (IV. 15). 3; 91 (IV. 17). 1. *istic* (Madvig) is unconvincing since C. has just said that he does not know where Atticus will be—or, taken vaguely as 'in the East', superfluous. *is* might seem better away; but cf. *Verr.* I. 21 *cum praetores designati sortirentur et M. Metello obtigisset ut is* (sc. *M. Metellus*) *de pecuniis repetundis quaereret*; *Fam.* III. 6. 1 *quaesivi ex Phania, cuius mihi videbar et fidelitatem erga te perspexisse et nosse locum quem apud te is teneret*; Plaut. *Mil.* 134 *nam et venit et is in proximo hic devortitur.* I find it less natural to take *is* with Sternkopf emphatically, '*he* (as distinct from other travellers to Epirus not so likely to see Atticus)'.

13 **Eutychide** Cf. 90 (IV. 15). 1. Nothing otherwise known.

90 (IV. 15)

1, 1 Eutychide Cf. 89 (IV. 16). 9. Atticus had chosen to give him his own former praenomen Titus instead of his present adoptive praenomen Quintus: cf. 67 (III. 22). 3 n.

3 **tua sqq.** *tuam erga me benevolentiam cognosse* ⟨*et*⟩ *suam* sqq. Bücheler. But C. is not likely to have cared (or admitted that he cared) to impress a slave with the warmth of Atticus' sentiments towards himself; and *suam... fuisse* becomes ambiguous (with *cognosse* or with *gratum est*?).

4 **dolore** The exile.

2, 5 praetor Clodius C. Claudius Ap. f. Pulcher (cf. 62 (III. 17). 1 n.), Pr. 56 and at this time Proconsul in Asia (cf. Broughton, p. 218). After his return he was prosecuted and probably condemned *de repetundis* (cf. *Fam.* VIII. 8. 2); he may have remained in exile until 43 (cf. *Fam.* XI. 22. 1). C. was on bad terms with him, as with his brothers (cf. Ascon. 26. 2). Atticus would doubtless have to do with him as governor on his business trip to the province (cf. 13 (I. 13). 1; 21 (II. 1). 12; 46 (III. 1)). Originally provincial governors had all been Praetors and were still commonly so called (cf. Mommsen, *St.* II, p. 240), even consular Proconsuls like Bibulus (cf. *Fam.* II. 17. 6) and C. himself (cf. 114 (V. 21). 11). The whole passage is of course ironical. Nothing could be more wanton than to substitute *rhetor* (Bosius), i.e. Sex. Clodius, the Sicilian rhetorician of *Phil.* II. 43, etc.

7 **Pituanius** Probably on the Proconsul's staff. The astrologer L. Pituanius who was thrown from the Tarpeian Rock in A.D. 16 (Tac. *Ann.* II. 32. 3) may have been a descendant. *Graecis litteris* is no doubt facetious, Pituanius being concerned with business documents (perhaps with his chief's extortions). I hardly credit Sjögren's idea that *Pituanios* of the MSS is a 'Grecism' on C.'s part (cf. 21 (II. 1). 5 n. *novus patronus*).

homo esse 'To be a decent fellow', with the ordinary allowance of humanity and proper feeling; so in *Fam.* XV. 17. 3 *quod se in his malis hominem praebuit*, XI. 29. 2 *quod tuum iudicium nisi mea summa benevolentia erga te... comprobaro, ipse me hominem non putabo* et sim. For a different nuance cf. 353 (XIII. 52). 2 *homines visi sumus.*

3, 1 dedi The letter, not having been delivered, has not survived. ἡμερολεγδόν suggests something in the style of 75 (IV. 3).

2–3 **ut conicio** *Secl.* Ernesti: but cf. 42 (II. 22). 6 n.

4, 1 Sufenas M. Nonius Sex. f. Sufenas, whose father was Sulla's nephew and Praetor under his Dictatorship, was C. Cato's colleague as Tribune in 56 and (no doubt) his collaborator in blocking the elections in the 'triumviral' interest: cf. Dio, XXXIX. 27. 3. He is probably not the governor mentioned in 115 (VI. 1). 13 (see *ad loc.*). He joined Pompey in the Civil War: cf. Plut. *Cic.* 38.

2 **Cato** Cf. 89 (IV. 16). 5 n. He was defended by Scaurus (Ascon. 18. 16) and prosecuted by Asinius Pollio and Licinius Calvus (Sen. *Contr.* VII. 4. 7, unless this refers to his previous trial under the lex Junia et Licinia.

Procilius Seemingly prosecuted like Cato and Sufenas in connection with the disturbances in 56; that he too was Tribune in that year (cf. Broughton, p. 209) is a probable inference, but no more. Rightly regarded, the remark in Q.*Fr.* II. 7. 1 of May 56, *Id. Mai. senatus frequens divinus fuit in supplicatione Gabinio deneganda. adiurat Procilius hoc nemini accidisse*, also suggests a henchman of the 'Triumvirs'. C. would hardly have thought Procilius' comment worth repeating if it had not come from the other side. His tone here as in 89 (IV. 16). 5 is clearly unsympathetic.

6 **XXVIII** *XXIX* in T.–P.'s text is a survival from Tyrrell's first edition, where it may not have been just an error, as Müller supposed. 51 seems to have been the normal number for a jury in murder cases: cf. Mommsen, *Str.* p. 218. But one of the jurors may have put in a blank vote: cf. *ibid.* p. 423.

7 **lacrimans** Tears in a Roman peroration were well within the rules: cf. Wilkins on *de Orat.* II. 190. *la-* falling out, *-crimans* or *-crumans* became *criminans.* That reading would imply that Clodius prosecuted. But these trials were political, aimed against the dynasts, and the alignment of personalities is otherwise consistent: for the prosecution Lucretius Vespillo, Calvus, Pollio (at this time probably in the optimate camp: cf. Catull. 12. 6); for the defence, Scaurus and (under cover: cf. 89 (IV. 16). 5 n.) Drusus. For Clodius to prosecute an associate of his old ally C. Cato (cf. Q.*Fr.* II. 1. 2) for 'pro-triumviral' activities in 56 makes no political sense. Further *Publius...moverat* reads naturally as an explanation of the size of the not-guilty vote. And why put in *criminans* at all? What else should a prosecutor be doing? ⟨*me*⟩ *criminans* (Madvig) escapes these objections. But C. does not seem to have been relevant to the case, and it would be unlike him to make a point against himself in so casual a fashion. Nor would he have gone on to explain that he had kept silent because Tullia (*pusilla*) was afraid that his interference might annoy Clodius if Clodius had just launched a powerful attack against him.

8 **verbum nullum** Cf. 18 (I. 18). 6 *Crassus verbum nullum contra gratiam.*

10 **offenderet** *offenderem* Manutius; but *verba facere* is not difficult to understand. Since anything that C. had to say would clearly have been in Procilius' disfavour, Tullia's apprehensions make a further point against Clodius' supposed role of prosecutor.

5, 1 **Reatini...duxerunt** Cf. *Scaur.* 27 *ego nuper, cum Reatini, qui essent in fide mea, me suam publicam causam de Velini fluminibus et cuniculis apud hos consules agere voluissent, non existimavi me neque dignitati praefecturae* *gravissimae neque fidei meae satis esse facturum, nisi me causam illam non solum homines sed etiam locus ipse lacusque docuisset.*

2 **causam** The ground of the dispute seems to have been as follows: by a cutting through the mountains Lake Velinus had been connected with the River Nar to the north. The author of the operation was M'. Curius Dentatus, Censor in 272 and builder of the second Roman aqueduct. Before the cutting was made, the only outlet for the lake waters in flood would have been to the south, into the fertile plain of Rosia or Rosea (*Rosea rura*; modern le Roscie), *sumen Italiae* as Caesar Vopiscus called it (Varro, *R.R.* I. 7. 10; Plin. *N.H.* XVII. 32), which belonged to Reate; this was thus relieved of inundation, though remaining sufficiently moist. The Interamnates complained, perhaps as a result of recent flood damage, that the nuisance had been diverted on to their territory. C. won his case, for in A.D. 15 (Tac. *Ann.* I. 79) we have the Reatines protesting against a plan to block Lake Velinus at its confluence with the Nar (i.e. the cutting) on the ground that the land adjacent (i.e. the plain of Rosea) would be flooded by the lake. But this time the blockage was not proposed by Interamna, but as part of a scheme to reduce floods in the capital by diverting Tiber tributaries. The Interamnates themselves were against the scheme, which included canalization of the Nar.

consules Cf. *Scaur.* 27 *apud hos consules*, where *hos* guarantees the plural. As this speech was delivered hardly more than a month later (cf. Ascon. 18. 4) a genuine discrepancy is unlikely, whereas *coss.* may easily have become *cos.* (cf. §7 *consulem* for *consules* and *consulis* for *consulum*). In the opening of Book III of Varro's *de Re Rustica* (III. 2. 3) Ap. Claudius reminds Axius of a visit at his Reatine villa paid a few days previously in connection with the Reate–Interamna dispute; but I doubt whether the assumption is correct that C. and Varro are referring to the same occasion. C. says nothing about having Appius as his fellow-guest, nor does *ut agerem* necessarily imply that the case was heard locally; C. went to collect information, as he says himself in *Scaur. l.c.* If the hearing had already taken place he would surely have told Atticus the result. Indeed it seems more than questionable whether Book III of the *de Re Rustica* is rightly set in the year 54. Nothing in the dialogue indicates that Appius was Consul at the time. On the contrary Varro introduces him as *augurem sedentem in subselliis ut consuli, si quid usus poposcisset, esset praesto*. The occasion was the elections, at which Varro and Axius had just voted, to the Curule Aedileship (a Consul would not have presided over elections to the Plebeian Aedileship: cf. Mommsen, *St.* II, p. 483). These, according to the established order (*ibid.* I, p. 580, n. 2), followed those of the superior magistrates. But no Consuls were elected in 54, therefore presumably no Praetors or Curule Aediles.

3 **qua** C. had not been asked to conduct the case *because* the lake waters

were flowing into the Nar, as they had been doing presumably for over two centuries. He does not trouble to tell Atticus just what had provoked the Interamnates to protest in 54; no doubt their grievance was of long standing.

4 **Narem** There is no support for the neuter form: cf. Neue–Wagener, *Formenlehre*³, IV, p. 955.

modice Cf. Colum. *R.R.* III. 6. 3 *loco modice umido, non uliginoso.*

5 **Axio** Cf. 12 (I. 12). 1 n. He had a very smart (*perpolita*) villa near the lake (Varro, *R.R.* III. 2. 3).

Septem Aquas A group of lakes to the north of the Plain, perhaps to be identified with Virgil's *fontes Velini* (*Aen.* VII. 517) and including the peripatetic Fons Neminie (Neminiae? cf. Plin. *N.H.* II. 230: also *ibid.* XXXI. 12 *Cicero in 'admirandis' posuit Reatinis tantum paludibus ungulas iumentorum indurari*). A *pagus* was named after them (cf. Philipp, *RE.* IIA. 1550. 12). I see no particular advantage in writing *qui* (Wesenberg) for *quin*. Such a country excursion would be something quite out of C.'s ordinary habits, from what we know of them.

6, 1 **Fontei** Identity uncertain, but probably not the governor of Gallia Narbonensis whom C. defended *c.* 69: cf. 2 (I. 6). 1 n. Broughton (p. 221) suggests that he was Praetor Urbanus this year and that C. returned to Rome in order to attend the *ludi Apollinares* under his presidency.

spectatum Better than ⟨*in*⟩ *spectaculum* (Malaespina): cf. *Sest.* 126 *via...qua spectatum ille veniebat.*

3 **Antiphonti** Otherwise unknown. Arbuscula is mentioned by Horace (*Sat.* I. 10. 77).

operam Sc. *dedi.*

4 **manu missus** In confidence of success. Successful performers were sometimes emancipated by popular demand: cf. Suet. *Tib.* 47; Dio, LVII. 11. 6.

palmam Wreaths, sometimes of gold, were awarded: cf. Fiebiger, *RE.* IV. 1642. 66.

5 **nihil tam** Best left with Bosius as an aposiopesis.

6 **haec...habeto** Cf. *Fam.* VII. 25. 2 *secreto hoc audi, tecum habeto, ne Apellae quidem...dixeris.*

Andromacha The role rather than the play: cf. Ov. *Rem.* 383 *quis ferat Andromaches peragentem Thaïda partis? / peccet, in Andromache Thaïda quisquis agat.*

maior Physically, of course; but surely there is a *double entendre*: cf. Sen. *Contr.* III, praef. 3 ⟨*auditus tamen*⟩ *longe maior erat quam lectus*; Quint. *Inst.* X. 1. 74 *in historia praedictis maior.* Normally *parem habuit neminem* would mean 'nobody else was so good' or 'so big', but the context here implies the opposite, which is equally to be found in the words, literally interpreted. Here for once the jest stays exactly the same in English.

9 **venatio** The beast-fight that ordinarily followed the games: cf. 411 (XVI. 4). 1.

7, 1 ardet Cf. *Q.Fr.* II. 15. 4 *ambitus redit immanis...res ardet invidia*; also Plin. *N.H.* praef. 9 *flagrantibus comitiis* (perhaps suggested by C.'s phrase; see the context); Tac. *Ann.* XIV. 28. 1 *comitia...quod acriore ambitu exarserant.* The whole passage is paraphrased in *Q.Fr.* II. 15. 4, written about the same time.

σῆμα κ.τ.λ. *Il.* XXIII. 326.

2 **triente...bessibus** Interest reckoned by the month.

4 **Domitium** So usually, but *Domitius Calvinus* in *Q.Fr.* III. 4. 1 and *Calvino* in 91 (IV. 17). 3. Here the context precludes any confusion with Ahenobarbus. The *pactio* stands revealed in 91 (IV. 17). 2.

7 **fronte an mente** Cf. *Planc.* 16 *tabella quae frontis aperit hominum, mentis tegit.* Similarly 103 (V. 10). 3 *fronte...intimis sensibus*; *Fin.* II. 77 *ut in fronte ostentatio sit, intus veritas occultetur*; Pers. V. 116 *fronte politus / astutam vapido servas in pectore vulpem*; Tac. *Agr.* 39. 1 *fronte laetus, pectore anxius.*

11–12 **HS quingena...id** Cf. K.–S. I, pp. 62 f.

8, 1 comitia Sc. *tribunicia.* There is no need to add the word, as Müller proposed.

4 **omnes leges** Cf. *Q.Fr.* II. 15. 4 *quae quidem comitia si gratuita fuerint, ut putantur, plus unus Cato potuerit quam omnes leges omnesque iudices.*

9, 1 Messius Like Drusus and Scaurus, and later Gabinius and Vatinius, a 'triumviral' figure, but in his case C. had a personal obligation to repay: cf. 73 (IV. 1). 7 n. Calvus also defended Messius, but not necessarily in this trial: cf. Sen. *Contr.* VII. 4. 8 *pro Messio tunc tertio causam dicente.*

2 **legarat** Appius as presiding magistrate in the Senate had nominated him, no doubt at Caesar's wish. That was normal constitutional form, though at this period a provincial governor in practice appointed his own staff: cf. Mommsen, *St.* II, pp. 677 ff.

Servilius The younger Isauricus, now Praetor: cf. 19 (I. 19). 9 n. His alliance with Caesar was evidently yet to come.

tribus Messius will have been prosecuted under the lex Licinia *de sodaliciis* of 55. It gave the prosecution the right to choose four tribes from which the jury had to be selected; the accused could reject one of these: cf. *Planc.* 36 ff. The charges probably concerned Messius' election as Aedile in 55, though the scope of the law is not accurately known: cf. Mommsen, *Str.* p. 872, n. 2. According to Dio (XXXIX. 32. 2) the elections to the Curule Aedileship for 55, which did not take place until after the beginning of that year, led to bloodshed. The Plebeian Aediles on the other hand had been elected in 56, probably at the normal time; certainly, therefore, prior to the passage of the lex Licinia. Even though this may have been made retrospective, like Pompey's bribery law in 52, the chances seem strong that

Messius was Curule Aedile in 55 (and not Nonius Struma: see Broughton, p. 283). Broughton makes him Plebeian Aedile because he celebrated the Floralia. But the evidence that only Plebeian Aediles did so (cf. L. R. Taylor, *Amer. Journ. Phil.* 60 (1939), p. 195) is not conclusive; there may even have been some alternation. Messius appears to have been acquitted in all his trials: cf. 161 D (VIII. 11 D). 2.

3–4 **pugnatur...satis** There is no warrant for such renderings as 'It is a sharp struggle, however it is getting on fairly well'. *sat(is) agere*, otherwise confined to pre-Ciceronian and post-Augustan writing, means 'to have one's hands full', 'be in trouble',—no doubt a colloquialism at this period. Both verbs have to be understood from the writer's point of view.

4 **Drusum** Cf. 89 (IV. 16). 5 n.

Scaurum His trial *de repetundis* ended in acquittal on 2 September (Ascon. 18. 3).

5 **gloriosi** Drusus and Scaurus bore names which their fathers (adoptive in Drusus' case) had made famous.

6–7 **in quibus...laborabit** A magistrate could not be brought to trial while in office, and a charge already brought against him was held over until the end of his term: cf. Mommsen, *Str.* p. 353. According to Paulus (*Dig.* III. 3. 54) a magistrate due to enter office (*qui magistratum initurus est*) was likewise immune. This must, I think, refer to all *designati* and not, as Mommsen seems to hold, to those due to take office within a certain period. It is true that prosecutions of *designati* sometimes took place, but I know of none which did not concern alleged misconduct in the election itself; such cases, it may be suggested, constituted a recognized exception within the rule.

10, 5 **roges...venias** Victorius' conjecture *roges et hortere ut quam primum veniat* used to be the vulgate before Sjögren, following Lehmann, returned to the MSS. It was altogether misguided. C. assumes that Dionysius' movements depend on those of his *patronus*.

91 (IV. 17)

1, 1 **non oblitum** *nunc* (Pius) is tempting. However, *puto te existimare* vel sim. ordinarily refers to something which the writer feels his correspondent *ought* to think in justice to himself: cf. 198 (x. 7). 3 *hoc autem vitium puto te existimare non ⟨a⟩ nostra indulgentia...profectum*; *Fam.* II. 16. 3 *credas hoc mihi velim, quod puto te existimare, me* sqq., XIII. 7. 4 *hoc me non sine causa facere...etsi te existimare arbitror, tamen mihi adfirmanti credas velim* sqq., XV. 4. 13 *cum, quem ad modum te existimare arbitror, non difficillime consequi possem*, XII. 12. 1 (Cassius) *quod te primum existimare putabam nos oppressa re publica quiescere non posse*; Q.Fr. III. 4. 2 *non existimo te putare id mihi susci-*

piendum fuisse, III. 5. 2 *puto enim te existimaturum a me illos libros non sine aliquo meo stomacho esse relictos.*

2–3 **cum...video** *quoniam* is often conjectured in such cases (here by Orelli) to avoid what Madvig on *Fin.* I. 10 calls 'apertus soloecismus': but cf. 264 (XII. 25). 1 *cum praesertim necesse erit*; 371 A (XIV. 17 A). 3 *gratulor tibi, cum tantum vales apud Dolabellam*; 395 (XV. 18). 2 *cum intellego te distentissimum esse*; *Fam.* VII. 15. 2 *cum vero in C. Mati...familiaritatem venisti, non dici potest quam valde gaudeam*; K.–S. II, pp. 348 f.

3 **in Epirum** Sc. *proficiscenti*. It seems unnecessary to follow Wesenberg in omitting the last *neque*. No traveller to Epirus or to Athens or to Asia (the natural places to find Atticus), no traveller in a word whosoever (cf. *post Red.* 10 *neque me neque quemquam fefellit* et sim.), could be trusted with a letter unless he was going to meet Atticus, wherever he might be found.

5 **eae sunt** *eae* has been added in various positions. The MSS present a number of examples of omission, such as 134 (VII. 11). 5 *vult enim me Pompeius esse quem tota...ora habeat* ἐπίσκοπον: cf. Sjögren, *Comm. Tull.* pp. 163 f. But this seems particularly awkward.

8 **ne quid quo** Constans adopts Starker's *ne quid umquam*, but *quo* should be retained, as in several conjectures of which this is the best: cf. 104 (V. 11). 5 *nec...habeo quod adhuc quem accusem meorum*; 115 (VI. 1). 20 *num quid de quo inaudisti?*; 224 (XI. 13). 5 *si quid erit quod ad quos scribendum meo nomine putes*; 251 (XII. 14). 3 *non quo proficiam quid.*

2, 3 |XXXX| I.e. *quadragies*. 40,000 sesterces (*quadragena*) is an absurdly small sum in this context; it is enough to point to the 500,000 which each of the tribunician candidates deposited with Cato (cf. 90 (IV. 15). 7). Moreover Appian in what seems to be a reference to the affair mentions 800 talents, which would correspond to HS 19,200,000: cf. *B.C.* II. 19 ὤφθη δέ που καὶ μεσεγγύημα ταλάντων ὀκτακοσίων ὑπὲρ τῆς ἐπωνύμου γενόμενον ἀρχῆς. Manutius and others read *quadringena* (400,000); but even this falls far short of Appian's figure and of what is plausible. Eduard Meyer (*Caesars Monarchie*, p. 195, n. 2) makes it 40,000,000 (so Billerbeck translated, and so too Münzer, *RE.* XV. 614. 3). That is near enough to Appian, if we suppose that each of the candidates was responsible for 20,000,000; but it means reading *quadringenties* or |*CCCC*|, πρὸς τὸ καρτερόν. Besides, this amount, roughly that of D. Brutus' entire fortune in 44 (*Fam.* XI. 10. 5), does seem excessive. The remaining possibility is 4,000,000, and that is assumed by Gelzer (*Röm. Nobilität*, p. 99, n. 6), who tells us to understand *HS quadragena* as 40 × 100,000. Appian, he says, has evidently put together both numbers in C. ('die beiden Zahlen bei Cicero zusammengezählt'). I suppose he means that Appian regarded each candidate as responsible for 8,000,000—4,000,000 for each Consul. We thus get a total of 16,000,000—not quite the same as

19,200,000, but near enough perhaps. Anyhow 4,000,000 is a reasonable sum, and the corruption is readily explained: $\overline{|XXXX|}$ became $\overline{XXXX}$. Conversely in 40 (II. 20). 6 *centies* ($\overline{|C|}$) has replaced *centena* ($\overline{C}$).

6 **ornandis** Presumably the Consuls had met with difficulties in obtaining this in the ordinary way. In the end Appius went to Cilicia without any *lex curiata*, or at any rate so proposed (cf. 92 (IV. 18). 4; *Q.Fr.* III. 2. 3; *Fam.* I. 9. 25), while Ahenobarbus did not take a province at all.

7–8 **cum...fuisset** On the date mentioned (presumably) in the *pactio*. I do not think it possible to translate 'though there had never been a meeting about it at all'.

8 **nominibus et perscriptionibus** 'By book entries and orders for payment', according to Roby, *Rom. Private Law*, II, p. 295. The latter would seem premature. It is difficult to be positive, but I should prefer to take *nominibus* as the names of the parties (so below) and *perscriptionibus* as entries of the particulars (cf. Wilkins on *de Orat.* I. 250).

9 **multorum** Besides the four principals it may be supposed that the Augurs and Consulars would record the *pactio*, or rather perhaps their individual arrangements with the candidates, in their ledgers.

per tabulas Cf. *Rosc. Com.* I *qui per tabulas hominis honesti pecuniam expensam tulerunt.* The vulgate *tabulis* is diplomatically inferior and doubtful Latin.

10 **nominibus inductis** Presumably the names of the Augurs and Consulars and of witnesses. There would be no object in cancelling those of the principals.

auctore Pompeio It seems strange that Pompey should have wished to expose Appius, whose daughter had married his son; but no doubt there were many wheels within wheels. Note the reference to Caesar's displeasure in §3.

11 **idem** Cf. 16 (I. 16). 11 n. *iidem*. The common interpretation, 'cela laisse Appius parfaitement impassible', suits Vatinius' reference to his *os* in *Fam.* V. 10. 2 but is not so consonant with usage. *iacebat* likewise refers to loss of reputation: cf. 115 (VI. 1). 18 *num igitur iacet Theophrastus...?*; 142 (VII. 18). I *quae quidem ille si repudiarit, iacebit.* Ahenobarbus is Lentulus Spinther's anonymous enemy in *Fam.* I. 9. 2 (December 54): *ea est enim conatus quibus patefactis nullam sibi in posterum non modo dignitatis sed ne libertatis quidem partem reliquit* (cf. *Philol.* 105 (1961), pp. 73 f.).

3, 2 **refrixerat** Cf. *Q.Fr.* III. 2. 3 (9 October) *Domitius cum Messalla certus esse videbatur; Scaurus refrixerat,* III. 6. 3 (end of November) *hic* (i.e. *Romae*) *quidem friget* (sc. *Memmius*). Calvinus does not seem to have been damaged by the scandal, which was apparently regarded as discrediting the Consuls more than the candidates (cf. §2 *consules flagrant infamia*).

2 **nunc** The letters *cociace* which follow in the MSS probably belong elsewhere: cf. 93 (IV. 19). 1 n. Mommsen made *hoc* (= *hac re*) *iacet* and Reid *totus iacet* out of them, but neither fits very happily.

4 **eius Domitius competitor** 'Collocatio verborum, qualis est Plaut. Capt. 875 *tuom Stalagmum servom*, Cic. de n.d. III 15[39], *eius Palaemonem filium*, in voce *competitor* vix probabilis', Sjögren: cf. his discussion in *Glotta*, 10 (1919), pp. 23 ff. In C. as in Plautus the person named in such cases is usually a child or slave: but cf. Plaut. *Merc.* 690 *tuam Alcumenam paelicem, Iuno mea*; Cic. *N.D.* III. 48 *huius Absyrto fratri*. The evidence hardly countenances dogma.

liberalis Apparently in distributing largesse: cf. §4.

5 **at** The particle here seems to have no connective force (cf. 44 (II. 24). 1 n.), so editors are probably right to prefer *at*; for consistency's sake the same choice will have to be made in 353 (XIII. 52). 1 where the MSS have *ac mihi Barba Cassius subvenit*.

6 **decernit** Historic present.

tacitum iudicium Plutarch (*Cat. Min.* 44) apparently refers to this episode (despite discrepancies) when he says that because of bribery scandals Cato persuaded the Senate to decree that *designati* should be obliged to give an account of their conduct before a sworn tribunal even if no one came forward to accuse them. It would thus seem that the absence of a prosecutor constituted the main feature of a *tacitum iudicium*, concerning which we have no other evidence. Such a trial, lacking the ordinary preliminaries of *postulatio*, was perhaps felt to be 'unofficial': cf. *tacitae indutiae*, 'an undeclared truce' (Liv. II. 18. 11, etc.).

ab iis consiliis sqq. If *omnibus* is dative, we must suppose that various tribunals had already been set up to try such cases of *ambitus* as might arise. But *omnibus* is obscure and the plural *consiliis* hard to explain. There was a standing *quaestio de ambitu*: cf. W. Kunkel, *RE.* XXIV. 744 f. Sternkopf (*Hermes* 40 (1905), p. 31) proposed *consiliis* ⟨*quae essent ex consiliis*⟩ *quae erant omnibus sortita*—the special tribunals were to be made up from all those already constituted for particular cases (cf. Mommsen, *ibid.* pp. 211 f.). But *iudicium* is against assuming more than one special tribunal. It might be better to change *sortita* to *sortito* (adverb); 'by all the standing tribunals, by lot', i.e. by a jury chosen by lot from all the standing tribunals.

8–9 **Opimius, Veiento, Rantius** None of these is certainly identifiable. Opimius may be the M. Opimius (probably grandson of the Consul of 121) who held a cavalry command under Metellus Scipio in 48 (Caes. *B.C.* III. 38. 4). A Veiento, probably a *legatus*, was left in charge of Syria by Bibulus in 50 (126 (VII. 3). 5). The Fabricius Veiento of Nero's time (cf. Tac. *Ann.* XIV. 50) may have been a descendant; but H. Gundel's citation (*RE.* VIII A. 590. 65) 'Cic. *Fam.* II. 15. 4 *sunt enim e Fabriciis Veientones*' is a corruption of Orelli's index. There is evidence for the cognomen also in the Perperna gens (cf. Münzer, *RE.* XIX. 897. 31). The nomen Rantius occurs in a republican inscription (*C.I.L.* I. 1624).

10 **res cecidit** I.e. *in praesentia.* Nothing further could be done without legislation, which the Senate proceeded to set in motion. *res cedit,* usually interpreted as *tractatur res* (cf. 40 (II. 20). 2), would rather mean 'the thing goes forward': cf. Suet. *Nero,* 33. 2 *cum opinione tardius cederet* (sc. *venenum*); *Thes.* III. 732. 74. *res cecidit* is also, so far as I know, a unique phrase, but can be supported by analogy: cf. *ibid.* III. 26. 44.

dilata By a new decree, now that the recalcitrance of the jurors had made legislation necessary.

11 **Terentius** An otherwise unknown Tribune.

12 **levi bracchio** Cf. 21 (II. 1). 6 *molli bracchio.*

13 **detulerunt** Cf. *referretur* below. But Müller points to 16 (I. 16). 5 *defertur res ad senatum*; 44 (II. 24). 2 *res delata ad senatum est* et sim.

Abdera The proverbial home of blockheads or crack-brains (perhaps at this period the latter rather than the former): cf. 130 (VII. 7). 4 *id est* 'Αβδηριτικόν; Otto, *Sprichwörter,* p. 1.

non tacente me C. does not mean that he 'contributed his share of noise' (no noise is implied), but that he protested against the absurdity of the Senate's behaviour (so Manutius): cf. *Mur.* 48 *atque ex omnibus illa plaga est iniecta petitioni tuae non tacente me maxima.*

16 **si quis** So Sjögren for the vulgate *si qui*: cf. *Fam.* VIII. 8. 6–8.

17 **leviter** 'Without pressure.'

19 **c.** The variation *censuerunt / cens(u)ere* in the MSS is covered by my hypothesis that Cicero used the customary abbreviation. He may well indeed have written *d. e. r. i. c. comitia p. q. t. haberi*: cf. *Fam.* VIII. 8. 5 f. The ritual formula (cf. Mommsen, *St.* III, p. 1008) emphasizes the absurdity of the anti-climax. *haberi* is regular: cf. Sjögren, *Comm. Tull.* p. 156.

4, 1 **absolutus** *de repetundis*: cf. 89 (IV. 16). 6.

1–2 **partem eius** *patrem eius,* as a joke παρὰ προσδοκίαν, seems to me too frigid for Ciceronian wit at its worst. *partem* cannot mean 'case', and the meaning 'cause' or 'side' would be inappropriate. As Sternkopf saw, it means 'part (of the case)': cf. *Rosc. Am.* 124 *si enim taceo, vel maximam partem relinquo*; *Mur.* 48 *cum...non partem aliquam agerem causae*; *Flacc.* 83 *quam partem causae tibi daremus?* Scaurus had been defended by a multiplicity of advocates (six, including Clodius, Hortensius, and Messalla Niger as well as C.: cf. Ascon. 20. 16), a practice of which C. strongly disapproved (cf. *de Orat.* II. 313; *Brut.* 207 f.). *eius* stands for *eius causae,* on the same principle as *Tigrane* in 53 (III. 8). 3 for *Tigranis negotio.*

2 **Scaevolam** Q. Mucius Scaevola, probably grandson (so Münzer, *RE.* XVI. 446. 26) rather than son of Scaevola the Augur (cf. 89 (IV. 16). 3 n.);

though, apart from his own Augurate (cf. 176 (IX. 9). 3), there is nothing to prove that he did not come from the elder branch of the family as son or grandson of Scaevola the Pontifex Maximus (Cos. 95). A friend of C. (and of Brutus: cf. *Brut.* 147), he was probably one of Q. Cicero's suite in Asia (cf. *Q.Fr.* I. 2. 13). After his Tribunate in 54 he was with Ap. Pulcher in Cilicia (*Fam.* III. 5. 5), but a Caesarian in the Civil War (cf. 176 (IX. 9). 3). On his fashionably improper verses cf. Plin. *Ep.* V. 3. 5; Münzer, *l.c.* 447. 18. The traditional view that his intervention at this time was hostile to Scaurus should be revised. There were family ties between them. Scaurus' wife Mucia was the daughter of Scaevola the Pontifex, while his sister had married M'. Acilius Glabrio (Cos. 67), whose mother Mucia was daughter of Scaevola the Augur and therefore probably the Tribune's aunt (cf. Münzer, *Röm. Adelsparteien*, pp. 275 ff.). Apart from such considerations, it is more natural to suppose that Scaevola's object was to hold up the elections, either to help Scaurus' candidature by giving him time to complete his largesses or for some reason of his own.

3–4 **quo...scripsi** Actually it was already 1 October, as C. presently notices: cf. 75 (IV. 3). 5 n. (*VIIII Kal.*). Unger (*Philol.* 51 (1892), pp. 214 f.), to suit his contention that the Roman day was always reckoned to start at midnight (cf. 75 (IV. 3). 4 n. *ante mediam noctem*), implausibly supposes that C. broke off his letter to get some sleep (presumably at *nundinarum fore*), and resumed it in the morning.

4 **satis fecerat** Cf. *Verr.* II. 5. 183 *cum...populo Romano satis factum... erit*; *Tusc.* II. 41 *vel domino satis facere vel populo* (of gladiators). Here, with reference to bribery, perhaps ironical. *domi suae* (cf. 16 (I. 16). 12 n. *cuius domi divisores habitarent*; also 93 (IV. 19). 1 *uno loco*) need not imply that Scaurus would have done this business in public but for Scaevola's 'obnuntiations' (cf. Constans, III, p. 96).

7–8 **nam...fore** Cf. 90 (IV. 15). 7. As a *faenerator* Atticus would be chagrined at having through his absence abroad missed this golden but fleeting opportunity. It turned out to be less fleeting than C. thought: cf. 93 (IV. 19). 2 n.

8 **sed** = δ' οὖν.

erat futurus 'Quid clarius quam librarium *fu* ex *fut.* occupasse?' (Wesenberg). *fuerat futurus* could only mean 'was to have taken place'.

10 **Ateium** The Antii recorded at this period are hardly more than shadows. A C. Antius (Restio?) was Tribune in 68 (cf. Broughton, pp. 138 f.); that speech of Sestius which gave Catullus a cold was *in Antium petitorem* (Catull. 44. 11); and Appian speaks of an Antius among the proscribed in 43 (*B.C.* IV. 40). C. Ateius Capito, on the other hand, Tribune in 55, was a staunch ally of Cato against the dynasts, and is linked by Dio with

Cato and Favonius as an opponent of the lex Trebonia (XXXIX. 35. 4f.; cf. XXXIX. 32. 3). On his harrying of Crassus, which earned him a censorial *nota* from Ap. Pulcher (*Div.* I. 29), cf. 87 (IV. 13). 2 n. He was a warm and constant friend of C. (*Fam.* XIII. 29. 2: cf. the letter to him, 407 C (XVI. 16 C)), and also, curiously enough, of Caesar (*Fam.* XIII. 29. 3, 6), whom he supported in the Civil War (*ibid.* §7) and under whose régime he flourished. But his persecution by Appius would in any case have inclined him Caesar's way in 49. The Senator L. Ateius Capito (*Fam.* VIII. 8. 5; *Q.Fr.* III. 1. 15), father of the celebrated jurist, was no doubt a relative. *ateium* would become *antium* all the more easily because the copyist had met the town Antium in two previous letters of Book IV.

5, 1 **Drusus** Cf. 89 (IV. 16). 5 n.

1–2 **non fecisse videntur** The customary legal formula: cf. *Verr.* II. 5. 14; *Acad.* II. 146.

2 **tres** Notice of prosecution was also given against Memmius by one Q. Acutius a few days later: cf. *Q.Fr.* III. 2. 3. From App. *B.C.* II. 24 it would seem that none of the candidates came to trial until 52, when Scaurus and Memmius were condemned. Messalla was acquitted in 51, but then condemned on a second charge. No prosecution of Domitius is recorded.

3 **Memmio** Not, so it is generally held, the candidate C. Memmius, but a cousin (?) of the same name who was Tribune this year. A consistent Senatorialist (cf. *Rab. Post.* 7), he was one of Gabinius' accusers *de ambitu* (*Q.Fr.* III. 1. 15, 2. 1) and also prosecuted Caesar's protégé Rabirius Postumus. But it is surely improbable that here and in *Q.Fr.* III. 2. 3 C. would have invited confusion by naming this Memmius without adding *tribuno pl.* to distinguish him, as he does in *Q.Fr.* III. 1. 15 (in III. 2. 1 no distinction was necessary, since Quintus would already know from the previous letter which of the two was prosecuting Gabinius). I therefore take it that the candidate is meant.

Q. Pompeio Rufo Grandson of Sulla's colleague in 88 and, through his mother Cornelia, of Sulla himself. As Tribune in 53–52 he promoted disorder and was arrested by the Senate in consequence (cf. Dio, XL. 45. 2); that he did so in furtherance of his *gentilis* Pompey's plans to become Dictator (cf. Miltner, *RE.* XXI. 2252. 43) is possible but unproved. He seems rather to have been associated with Clodius (cf. *Mil.* 90 *qui cum tantum ausus sit ustor pro mortuo, quid signifer pro vivo non esset ausurus?*), and was condemned on leaving office for his leading share in the riots which followed Clodius' death. His prosecutor was M. Caelius Rufus, who later came to his rescue when he was living in poverty at Bauli (cf. Caelius, *Fam.* VIII. 1. 4). An enemy of C.'s, he was rumoured to have murdered him in 51 (*ibid.*).

3 **Triario** Cf. 89 (IV. 16). 6 n. In the event it was he who gave notice to prosecute (*Q.Fr.* III. 2. 3).

4 **L. Caesare** L. Julius L. f. Caesar, son of the Consul of 64 (cf. 10 (I. 1). 2 n.) and a very distant relative of the future Dictator. Now a young man, his principal contribution to history was to be his role in the abortive peace negotiations of January 49. Unlike his father a staunch republican, he was killed in Africa, whether by or without Caesar's orders, in reprisal perhaps for cruelties which he had perpetrated upon some of Caesar's dependants (cf. Suet. *Iul.* 75. 3).

5 **libris** *de Oratore*.

6 **nihil reperio** The words that follow in the MSS, *nunc ut opinionem habeas rerum ferendum est*, have no tolerable sense here, either as they stand or as rewritten by Sternkopf ⟨*id*⟩ *nunc...reorum, ferendum est.* They must belong to the next letter. *in illis...reperio* makes a very satisfying conclusion to this one. We might indeed have expected some report of Gabinius' return, so possibly the end is missing. On the other hand C. did *not* reckon to give Atticus all the more important news (cf. 42 (II. 22) intr. note).

92 (IV. 18)

On 19 September A. Gabinius arrived outside Rome (*Q.Fr.* III. 1. 15) from his province of Syria and, having surrendered hopes of a Triumph, entered the city on the night of the 27th (*ibid.* 24). He had to face three prosecutions. The first, *de maiestate* for his high-handed restoration of Ptolemy Auletes, ended in his acquittal on or just before 23 October (cf. *Q.Fr.* III. 4. 1, 6); C. would have liked to prosecute, but restrained himself (*ibid.* III. 2. 2). Instead he found himself solicited by Pompey for a reconciliation with Gabinius, which he at first refused. The letter begins *in mediis rebus* after a lacuna. To start with *quaeris* (so Sjögren) is absurd, apart from the problem of *nunc...ferendum est.* How was Atticus to know without being told that *ut gesserim* refers to a conversation with Pompey?

1, 1 **rerum** This can hardly be sound. *opinio rerum* is an odd phrase, and *meam* (Purser) a desideratum: cf. 161 (VIII. 11). 4 *habes coniecturam meam* et sim.

ferendum est Cf. *Rosc. Am.* 145 *patior et ferendum puto*; Symm. *Ep.* IV. 50 *ferendum!*

2 **ille** Pompey.

4 **putabat** 'Said he felt' or 'seemed to feel': cf. *Propertiana*, p. 213.

5 **absolutus** Sc. *Gabinius* (cf. *Q.Fr.* III. 4. 1). Formal ambiguities of this sort are common enough where the context precludes misunderstanding.

πρῷρα πρύμνα A restoration not merely 'worth considering' but extremely probable (editors tend to prefer γοργεῖα (*sic*) γυμνά (Bosius) or other grotesques): cf. *Fam.* XVI. 24. 1 *mihi prora et puppis, ut Graecorum proverbium est, fuit a me tui dimittendi ut rationes nostras explicares.* In Greek sources the point of the saying varies (cf. Leutsch–Schneidewin, *Corp. Paroem. Graec.* II, p. 447), in C. it clearly means 'the alpha and omega' or, more loosely, 'the main thing': cf. Dio Chrysost. XXXVII. 36 ὑμεῖς γάρ ἐστε νῦν τὸ δὴ λεγόμενον πρῷρα καὶ πρύμνα τῆς Ἑλλάδος. T.–P. add *et* (better καί), but the asyndeton is of a common type: cf. Kühner–Gerth, *Gr. Gramm.* II, p. 346; K.–S. II, pp. 149 f.

6 **infantia** *Sensu etymologico*, as in *Orat.* III. 142; similarly *infans* in 182 (IX. 14). 2 and elsewhere. Translations like 'infantile', 'faiblesse enfantine', err.

L. Lentuli Cf. 44 (II. 24). 2 n.

7 **deinde** Here in place of the adversative particle which would normally follow *omnino* (= μέν)—cf. *Q.Fr.* III. 4. 1 below.

contentio...sordes Cf. *Q.Fr.* III. 3. 3 *Pompeius vehemens in iudicibus rogandis*, III. 4. 1 *omnino nihil accusatore Lentulo subscriptoribusque eius infantius, nihil illo consilio sordidius; sed tamen nisi incredibilis contentio preces Pompei, dictaturae etiam rumor plenus timoris fuisset, ipsi Lentulo non respondisset*; Dio, XXXIX. 62. 3.

9 **reliqua** There were two. In one, *de repetundis*, Gabinius was defended by C. himself (cf. *Rab. Post.* 19); but he lost the verdict and retired into exile. The other, *de ambitu* (cf. § 3 n.), may have lapsed, though this did not necessarily follow; Milo was found guilty *in absentia* of three offences, including *ambitus*, after his original conviction *de vi*: cf. Ascon. 54.

2, 1 **belle** Cf. 103 (V. 10). 3 *haec ipsa fero...belle.*

2 **me...amo** Cf. 187 (IX. 18). 1 *at ego me amavi.*

9 **dant** The subject is *multa*, not *ii*.

nec tamen...demigro I.e. 'I do not take an active part *against* the optimates'.

10 **est ad naturam** Cf. 52 (III. 7). 2 n. C. usually represents κατὰ φύσιν by *secundum naturam* or *e natura*. Seneca (from Epicurus) has *si ad naturam vives, numquam eris pauper: si ad opiniones, numquam eris dives* (*Ep.* 16. 7). Here 'nature' evidently means 'my nature', but it hardly seems necessary to add *meam* or *nostram* in the text, especially as the general meaning is not actually excluded.

14 **isti** The aforesaid envious persons, Hortensius and the like. Sjögren could keep *ista*.

pedibus trahantur Cf. *Fam.* VII. 32. 2 *trahantur per me pedibus omnes rei.* The origin of this expression is not wholly clear. Troublesome persons might be expelled from a house or the like *pedibus tracti*, as Odysseus was, in imagination, by the suitors (*Od.* XVI. 276): cf. Plaut. *Rud.* 660 *proripite hominem pedibus huc itidem quasi occisam suem*; Juv. V. 125 *duceris planta* *velut ictus ab Hercule Cacus / et ponere foris*; Suet. *Rhet.* 6 *pedibus e tribunali detractus est*; Liv. XXIV. 22. 9 *Dionysi tyranni vocis, qua pedibus tractum, non insidentem equo, relinquere tyrannidem dixit debere.* But even in these passages the idea of dragging off a dead body is not far away, and in C. more seems implied than mere ignominious removal. Perhaps understand *a carnifice*, comparing Housman on Juv. I. 157 *et latum media sulcum deducit harena* (also Quint. *Decl.* XV. 14 *dum carnifex trahit*, CCXLVII (Ritter, p. 15. 1) *non vulneratum cruentumque per ora populi traxit* (sc. *carnifex*), possibly Sall. *Cat.* 51. 34 *illi quibus Damasippi mors laetitiae fuerat paulo post ipsi trahebantur*—but this may be bracketed with Tac. *Ann.* IV. 70 *trahebatur damnatus*).

14 **ἐμφιλοσοφῆσαι** 14 (I. 14). 4 n. (*auditori*). συμφιλοσοφῆσαι (Wyttenbach) was a *Falschverbesserung*.

15 **animi** C. seems to forget that *stomachus* is a physical organ.

concalluit Cf. *Fam.* IX. 2. 3 λεληθότως *consuetudo diurna callum iam obduxit stomacho meo*, also 38 (II. 18). 4 *sed iam prorsus occallui.*

16 **delectent** C. can be happily indifferent to public affairs, but only if his private life is satisfactorily arranged, Atticus' presence being of prime importance. The vulgate goes *locus...concalluit; privata modo et domestica nos delectant. miram* sqq. What then becomes of *delectatio oratoria*?

18 **tam consentientibus sensibus** Cf. *Amic.* 65 *neque vero qui non isdem rebus movetur naturaque consentit aut fidus aut stabilis potest esse.* The jingle lightens the sentiment—a touch of *verecundia* (cf. 17 (I. 17). 7).

3, 1 **interregnum** This was duly reached, and lasted till the following July.

2 **dictaturae** For Pompey, naturally; cf. the contemporary letter to Quintus (III. 4. 1) cited above (§ 1). Hirrus, now Tribune elect, actually made the proposal in 53 with some support from his colleagues, but for the time being without success (cf. Broughton, p. 228).

4 **P. Sulla** The Elder (cf. 75 (IV. 3). 3 n.), himself condemned *de ambitu* in 66. He was supported by his son and two other relatives (cf. *Q.Fr.* III. 3. 2). If he had obtained a conviction he would have rehabilitated himself, replacing his victim in the Senate: cf. Mommsen, *Str.* p. 509. Indeed this may have happened, though if so Gabinius was condemned *in absentia*: cf. § 1 n.

5 **foris esset** Cf. *Pis.* 12 *egere, foris esse Gabinium.* The two contexts are supposed to point to the meaning 'be insolvent' for this otherwise unattested phrase, a meaning which is definitely excluded here. Even if Gabinius' own funds were running short, which after nearly four immensely lucrative years in Syria (cf. Dio, XXXIX. 55. 5) is hardly credible, he would still have had Pompey's limitless exchequer to draw upon. Dio (XXXIX. 63. 2) says that Gabinius spent less on his second trial, but only because he was confident of acquittal. In *Pis.* 12 there is doubtless a financial *implication*;

but the usual view makes a tautology, and editors change the text. In both passages a more general sense, 'out', 'done for', suits excellently. As origin (though this must remain doubtful on any interpretation) one might look to *foras ferri*, 'be carried out to burial': cf. Plaut. *Aul.* 156 *foras feratur*; Sen. *Ep.* 12. 3 (of an old janitor) '*quis est*' *inquam* '*iste decrepitus et merito ad ostium admotus? foras enim spectat.*' In English 'out of doors' used to mean 'lost', 'worthless', 'irrelevant'; cf. Dr Johnson's and later dictionaries, also the following from G. M. Trevelyan's *England under Queen Anne*, I, p. 140: '"There is not", Aglionby reported in April [1701], "a port between London and Leghorn where we can shelter ourselves in case of war and so refresh. Leghorn itself is but precarious now, so that the Turkey trade is quite out of doors."' Sulla brought his charge (before the acquittal *de maiestate*; cf. *Q.Fr.* III. 3. 2), expecting to find Gabinius easy meat; for even if he survived the first trial, his unpopularity was such that he could hardly remain a Roman citizen (cf. *ibid.* 3 *odio premitur omnium generum...quid futurum sit nescio; locum tamen illi in civitate non video*). A correct understanding of *foris esse* may perhaps help in a puzzle later on, 377 (XV. 1). 2 *Dolabellam spero domi esse.*

5 **contra dicente** He wanted to prosecute himself. So Caecilius Niger and C. contested the right to accuse Verres.

6 **Torquato** L. Manlius L. f. Torquatus (cf. *Q.Fr.* III. 3. 2), son of a Consular who had prosecuted and replaced P. Sulla in 66. They belonged to the elder of the two main branches of a patrician house which latterly, until its recent Consulship, had fallen away from its old glories. The hereditary feud with P. Sulla had already stimulated the younger Torquatus to two prosecutions, and his action in Gabinius' case was no doubt in part an attempt to put a spoke in Sulla's wheel. C. says (*ibid.*) that everyone was glad he failed, presumably because Sulla was considered the more reliable prosecutor. Later Praetor (not, I think, in 49), Torquatus perished with Metellus Scipio in 46. C. knew him well and gave him the Epicurean role in his *de Finibus*; Atticus perhaps even better (cf. Intr. p. 8). He is probably the bridegroom of Catullus' *Epithalamium*.

8 **M. Fulvius Nobilior** A Knight of that name is mentioned by Sallust (*Cat.* 17. 4) among the Catilinarians. The Nobiliores were a noble plebeian family, long decayed: cf. 75 (IV. 3). 3 n. (*Q. Flaccus*).

9 **ne respondent quidem** Sc. *citati*: cf. Varro *ap.* Gell. XI. 1. 4; Ascon. 56. 3 et sim. They let their cases go by default (less probably the phrase might refer to failure to reply to a prosecutor's speech, as in *Q.Fr.* III. 4. 1 *ipsi Lentulo non respondisset*). But there is a *double entendre*, and *urbani* is ironical. Failure to answer a question or summons or greeting is the reverse of polite: cf. *Tusc.* I. 17 '*quid? si te rogavero aliquid, nonne respondebis?*' '*superbum id quidem est*'; Mart. IV. 83. 3 *securus nullum resalutas, despicis omnes.*

C. has a similar play in two letters addressed to jurisconsults: *Fam.* I. 10 *te superbum esse dicunt quod nihil respondeas* and VII. 16. 3 *superbiam tuam accusant, quod negent te percontantibus respondere.* Not 'cum iudices cognoscendi molestia liberabant' (Boot).

4, 3 **Sopolidis** A celebrated portrait painter in Rome *M. Varronis iuventa* (Plin. *N.H.* XXXV. 147 f.), i.e. in the first quarter of the century. This pupil, otherwise unknown, was prosecuted under a law passed in 65 against false assumption of Roman citizenship. His words when he heard the verdict doubtless contain a reflection on the happier fate of his patron, just acquitted under the *lex maiestatis* (cf. *Cluent.* 97). A quantity of guesses have produced two plausible suggestions: that *res* should be *reus* (ς), and that the point is to be looked for in a pun on *Papia* and *Paphia.* The aspirate *ph* is no obstacle when it is remembered that *Aprilis* could be derived from *Aphrodite* (cf. Varro, *L.L.* VI. 33), *putere* from πύθεσθαι (cf. *ibid.* 96), and Κρόνος identified with Χρόνος (by C. in *Nat. Deor.* II. 64, with many earlier and later authorities: see A. S. Pease *ad loc.*). Perhaps Antiochus apostrophized the law: *pro lex maiestatis!* The Greek could be a quotation, as οὐ σοὶ μερὶς ἅμα Παφίῃ—the sort of remark that might have been made to Hera or Athene on Mt Ida. But we are never likely to know.

5 **Pomptinus** C.'s MSS give the name in a variety of forms, but this preponderates and has Dio's support (cf. Gundel, *RE.* XXI. 2421. 3)—whether nomen or cognomen is doubtful. C. Pomptinus, the only man found with it, was a soldierly personage, Crassus' Legate against the slaves in 71. As Praetor in 63 he and L. Flaccus arrested the Allobrogian envoys, as governor of the Narbonensis (62–59) he repressed their rebellious countrymen. He had been waiting five years outside the *pomerium* for this Triumph. Caesar's party is said to have been behind some of the obstruction (cf. Schol. Bob. (Stangl), p. 149), but the political background is untraceable. In 51–50 Pomptinus was C.'s Legate in Cilicia and largely responsible for his military successes.

7 **Q. Mucius** Cf. 91 (IV. 17). 4 n. and the parallel passage in *Q.Fr.* III. 4. 6 Ἄρη πνέων *Q. Scaevola.*

latum de imperio Since a commander holding *imperium* by *prorogatio* forfeited it as soon as he crossed the *pomerium*, special legislation was required to authorize its continuance for the day of the Triumph (cf. Mommsen, *St.* I, p. 132). In Pomptinus' case the necessary law had been carried at dawn, before the legal hour, by an old comrade in arms, the Praetor Ser. Galba (cf. Dio, XXXIX. 65. 2). *insulse* is not 'stupidly', since Galba's manœuvre succeeded, but 'tactlessly', a facetious way of saying that it rode roughshod over constitutional niceties.

10 **nihilum** The regular prose form, as in *Orat.* 233 *ne ad nihilum omnia recidant*; *Phil.* VII. 27 *ne ad nihilum recidere patiare.* So *venire ad nihilum* (*Fam.*

XI. 12. 1; *Tusc.* II. 5), *interire* (*occidere*) *in nihilum* (*Acad.* I. 27; *Div.* II. 37). Lucretius occasionally has *nil* after the preposition (I. 266, 544), often *nilum*.

11 **sine lege** Sc. *curiata*: cf. 91 (IV. 17). 2 n.

5, 2 **litteras** Perhaps a joint letter: cf. R. Stark, *Rhein. Mus.* 94 (1951), pp. 200 ff.

2 **datas...Oct.** A transcript from the letter(s); unless these words, misplaced in the MSS, are to be regarded with Stark as an interpolation. His article 'Interpolierte Daten in Ciceros Briefen' (*l.c.* pp. 180–202), like most *Interpolationsforschung*, promotes uneasiness here and there without carrying general conviction. *proxime* was conjectured by Sternkopf (*Hermes*, 40 (1905), p. 37) on the hypothesis that C. had received a number of letters of which those dated 25 October were the most recent; I cannot think that C. would so have expressed himself. *proximis* (Boot) is not impossible, but I rather suppose that Caesar and Quintus used *proximae* to distinguish the main island from the whole group, Pythias' νῆσοι Πρεττανικαί (cf. Hübner, *RE.* III. 859. 52): cf. Plin. *N.H.* IV. 102 *Albion ipsi nomen fuit, cum Britanniae vocarentur omnes de quibus mox paulo dicemus.*

3–4 **confecta...reportabant** The style is that of a military report; but Stark's imputation of irony seems gratuitous. Caesar might well write so (cf. his letter to Q. Pedius in 182 (IX. 14). 1) and C. be reproducing him. *nulla praeda*, followed by *imperata tamen pecunia*, is simply factual, not 'parodierend-nüchtern'.

4 **Q. Pilius** Q. Pilius Celer, brother (or other relative) of Atticus' wife. He followed Caesar in the Civil War and later earned C.'s displeasure by carrying letters from M. Brutus and C. Antonius to the Senate: cf. *ad Brut.* 5. 3.

93 (IV. 19)

This letter seems to be roughly contemporaneous with *Q.Fr.* III. 6 written not long after 23 November (§ 5), since both allude to Q. Cicero's winter quarters. C. had just had word of Atticus' return to Italy.

1, 4 **δέρρεις** Probably leather tarpaulins put up on a ship's sides for protection in rough weather: cf. Caes. *B.C.* III. 15. 4 and T.–P. *ad loc.*

6 **an** Vestorius lived neither in Rome nor in Apulia, but in Puteoli; therefore *an*, not *nam* or *num* (Pluygers).

7 **Latinus 'Αττικισμός** Playfully sarcastic. Vestorius' lack of culture prompted two other jokes to which commentators have done less than justice: see on 116 (VI. 2). 3 and 366 (XIV. 12). 3. Here I think 'Αττικισμός plays on Atticus' name—Vestorius might know no Greek but he was still an 'Atticist'.

8 **quin...invisis sqq.** Cf. Caelius, *Fam.* VIII. 14. 4 *curre per deos atque*

homines! et quam primum haec risum veni, legis Scantiniae iudicium apud Drusum fieri, Appium de tabulis et signis agere.

9 **putamina** On the one hand we need a noun to govern the genitive *rei publicae*; on the other, we have *puta*, a *vox nihili*. The deficiency is supplied by Wesenberg's *umbram* or *simulacrum* after *invisis*, but 'entia non sunt multiplicanda praeter necessitatem'; it is from the corrupt *puta* that the missing substantive should be sought. So Constans reasoned, but found nothing better than γῦπα. For *putamina* (to be found in an otherwise extravagant paper of T. G. Tucker's, *Hermath.* 15 (1909), p. 290) cf. Nonius, 232 *putamina non solum arborum sunt verum omnium rerum purgamenta. nam quicquid ex quacumque re proicitur putamen appellatur. Plautus in Captivis* [655] *nucleum amisi, reliqui pigneri putamina.* All of virtue in the free constitution has departed, leaving only the husks, the lifeless trappings.

9–10 **nummos...divisos** *-de* having been detached from *vide*, *nummos* became *nummis* and *divisos* followed suit.

10 **uno loco** To make the thing more obvious. It might even be done at a candidate's house: cf. 91 (IV. 17). 4.

olface An excellent proposal (cf. 91 (IV. 17). 3), particularly in view of 92 (IV. 18). 3 *est non nullus odor dictaturae.*

11 **iustitio** Cf. Mommsen, *St.* I, pp. 263 ff. Public business had apparently been suspended as a result of the breakdown of the political machine.

2, 2 **ludum** 'Amusement' or 'banter'.

Selicianae 'We hear of a usurer Selicius [cf. 12 (I. 12). 1 n.]; but how does that help us?' More, I think, than Tyrrell's *Felicianae*, a supposed allusion to the will of one Felix mentioned in *Q.Fr.* III. 7. 8 (December). The context demands a political reference; and the link between the political situation and Selicius is presented in 90 (IV. 15). 7. Rampant bribery had raised the monthly rate of interest from $\frac{1}{3}\%$ to $\frac{2}{3}\%$ in July. At the beginning of October the *faeneratores* were still reaping their harvest, and C. amuses himself with the picture of Atticus' chagrin at missing it (cf. 91 (IV. 17). 4). The elections were still to come, and a further rise would not be surprising. The amount of the rise depends on what is to be understood by *faenus unciarium* (*usurae unciae*), a complicated matter: see Klingmüller's discussion in *RE.* VI. 2189 ff., from which it appears that in the Twelve Tables the expression meant 10%, or perhaps $8\frac{1}{3}\%$, annually, and still meant the same when Sulla passed his *unciaria lex* of 88. According to the Greek fashion of monthly interest reckoning, which had now become usual, *uncia* should = $\frac{1}{12}$ of a *centesima* (1%) per month, i.e. 1% on a year of twelve months. But C. is evidently thinking of the old and, at any rate until very recently, established sense of *usurae unciae*. His *je m'en fichisme* covers both the collapse of the constitution and the financial profiteering which was one of its immediate results.

4 **tabula** Cf. *Off.* III. 89 *si tabulam de naufragio stultus arripuerit*, and, metaphorically, Val. Max. VIII. 1. 12 *tamquam fragmentum naufragii leve admodum genus defensionis amplexus*; Quint. *Decl.* XII. 23 *naufrago tabulam abstulisti, mortuo adplicas navem.*

7 **hibernam...optio** 'Ut sumeret quam vellet legionem in hiberna ducendam' (Manutius), the legions having already been assigned their quarters: cf. Sternkopf, *Jahrb. f. Philologie*, I (1897), p. 852. Suet. *Cal.* 8 *apud hibernas legiones* is compared.

8 **istorum** Cf. 92 (IV. 18). 2 n. *isti.*

9 **Pompeio** As proconsular governor of the Spains. Some difficulty about C.'s appointment had earlier arisen with Caesar: cf. *Q.Fr.* III. 1. 18.

10 **extra urbem** As Pompey's *legatus* C. would not be able to stay inside the *pomerium*. He intended to remain near Rome, as Pompey did.

15 **adventus** Lehmann compares *Fam.* XI. 29. 1 *magnum pondus accessit ad tollendam dubitationem iudicium et consilium tuum.*

[ad me] 'Hoc aut abundat aut mendosum est' (Manutius): cf. 76 (IV. 4) *quo die venies, utique cum tuis apud me sis.*

APPENDIX I

CICERO'S JOURNEY INTO EXILE

Early in 58 Clodius promulgated his bill *de capite civis Romani* interdicting from fire and water any man guilty of putting a Roman citizen to death without trial. Rather than face a fight for which he had been spoiling (cf. 38 (II. 18). 3 *aveo pugnare*—June 59), C. fled the capital in the middle of the night (Plut. *Cic.* 31, etc.). The bill *de capite* was passed, and later followed up by the promulgation of another, *de exsilio Ciceronis*, which mentioned C. by name and confiscated his property (for a recent summary of these events, see R. G. M. Nisbet, *In Pisonem*, pp. ix ff.).

The date of C.'s departure and his subsequent movements have raised argument. Here I need only refer to the articles of C. L. Smith (*Harvard Studies in Class. Phil.* VII (1896), pp. 65 ff.), W. Sternkopf (*Philol.* 59 (1900), pp. 272 ff.) and G. de Benedetti (*Historia*, III (1929), pp. 539 ff., 761 ff.).

It must be conceded to de Benedetti (p. 765) that we cannot determine precisely when C. left Rome. One of his last actions before doing so was to dedicate a picture of Minerva *custos urbis* on the Capitol (Plut. *Cic.* 31, etc.). That he did this on 19 March, Minerva's festival day, is a conjecture, in itself quite plausible (cf. *Fam.* XII. 25. 1) but by no means mandatory. On the other hand, de Benedetti's view that C. left early in March, which M. Gelzer (*RE.* VII A. 917. 21: 'im ersten Drittel des März') and Nisbet accept, is neither proved nor particularly probable. Our main clue is Plutarch's statement (*Caes.* 14) that Caesar did not leave to join his army until he had helped Clodius to drive C. out of Italy (more accurately, no doubt, out of Rome). Caesar, travelling at something like record speed, reached Geneva to encounter the Helvetii within eight days (Plut. *Caes.* 17; cf. Caes. *B.G.* I. 7. 1). That he arrived before 28 March, the day fixed by the Helvetii for their rendezvous on the Rhône (*ibid.* I. 6. 4) is probable, though not certain (Smith, p. 77). That he arrived much before is unlikely. The manner of his journey suggests that he had not a moment to lose. The Helvetian envoys sent to him on his arrival were told to return on 13 April (*B.G.* I. 7. 6); they would hardly have been asked to wait for more than a fortnight. De Benedetti argues that Caesar needed time to destroy the bridge over the Rhône and to raise troops; but it was to gain time that he put off his reply to the envoys (*ibid.*). Accordingly I return to the opinion that Cicero left Rome in the third week of March.

We do not know where he went in the first instance, but it would naturally be to one of his own properties. Tusculum is obviously too close to Rome. On the other hand he would hardly have gone as far away as Pompeii, for he must have been anxious for further news. Indeed he seems initially to have hoped for a speedy, triumphal recall (*Q.Fr.* I. 4. 4 *saepe triduo summa cum gloria dicebar esse rediturus*). Either then he chose his house at Antium, or if that also was too near Rome for safety, his villa at Formiae, or again his family home at Arpinum. The story of his dream *in villa quadam campi Atinatis* (*Div.* I. 59) has no relevance here, since the *campus Atinas* (Val di Diano; cf. Plin. *N.H.* II. 225) belonged not to Atina in Latium but to Atina in Lucania (Smith, p. 80; see below, p. 230). But all prospect of an early return was dashed by the announcement of Clodius' law *de exsilio Ciceronis*, and C. made up his mind to leave Italy. He probably wrote the first letter of *ad Atticum*, Book III, just before setting off.

How much time had passed since his departure? Probably not much. Otherwise we should expect some letters from C.'s first stopping-place prior to III. 1. And Clodius is not likely to have wasted any time before trampling conclusively upon his routed enemy. The fact that he soon afterwards issued an amendment to his *lex de exsilio* suggests that it was promulgated in a hurry.

It is certain and universally accepted that the next letter to Atticus chronologically is III. 3 and that it belongs to an early stage in the journey south. It announces a change of plan. C.'s original intention, all agree, was to head for Brundisium. He hoped that Atticus, who may have been intending to go to Epirus himself, would overtake him on the road (III. 1). Now, in a brief note, he announces a change of plan, having decided *multis de causis* to make for Vibo in the toe of Italy; there he hopes Atticus will join him, *statim*. The note was probably written at or near a parting of the ways, perhaps at Teanum Sidicinum, perhaps just south of Capua (a glance at the map will show the possibilities). At any rate after Capua Cicero deserted the Via Appia for the more westerly Via Popilia.

Our knowledge of his further progress depends on the next three letters to Atticus, III. 2, 4, and 5. III. 2 is dated 8 April, from Nares Lucanae, a station on the Via Popilia south of the Campano-Lucanian border (see below, p. 229). III. 4 (undated), as its contents attest, was written at Vibo, or soon after leaving Vibo *en route* for Brundisium. III. 5 bears (in the MSS) the date 6 April, Thurii. It is common ground that C. travelled down the Via Popilia to a farm near Vibo belonging to his friend and beneficiary Sicca. There he received a copy of an amended version of the *lex de exsilio*, denouncing penalties against persons harbouring him up to a certain distance (III. 4; Sternkopf, pp. 281 ff.). In consequence he started for Brundisium, which he reached on 17 April (III. 7).

The account accepted by editors is mainly due to Smith, whom Sternkopf in general follows. According to this, the date in III. 5 must be corrected from 6 April to 10 April, i.e. from *VIII Id.* to *IIII Id.*, as Corradus first proposed in the sixteenth century. Journeying down the Via Popilia C. reaches Nares Lucanae on 7 April, dispatching III. 2 next morning. He stays the night of the 8th *in campo Atinati*, arrives in Thurii on the evening of the 10th, and dispatches III. 5. Two more days account for the 85 (*sic*) Roman miles from Thurii to Vibo, where he arrives on the evening of the 12th. Remaining only one night at Sicca's, he starts for Brundisium on the 13th and reaches Tarentum, about 175 (*sic*) miles away, on the evening of the 16th (cf. III. 7).

This schedule involves so many improbabilities that rather than accept it I should give up the whole problem in despair.

(*a*) In III. 5 C. says that if Atticus is still in Rome it is too late for him to catch up. Yet C.'s object in going to Vibo was, as explained in III. 2, to wait on his friend Sicca's farm, where he could stay *pro suo iure*, for Atticus to join him: Sternkopf answers (p. 300) that C. already knew, perhaps from a letter of Terentia's, that the terms of Clodius' amendment would make it impossible for him to stay long at Vibo. In that case he might have been expected to wait at Thurii for definite news; and he would surely have been more explicit. What could he have thought Atticus would make of the sequence: 'Come to Vibo, where I can stay some time, and accompany me to Brundisium' (III. 2, in effect); and, two days later, 'If you're in Rome you cannot catch me up, but if you're on your way, we shall discuss what needs discussing when you overtake me'?

(*b*) 'Die Hauptstraße von Nares Lucanae nach Vibo führt über Thurii (115 M von Nares Luc., 85 von Vibo entfernt)': so Sternkopf, *l.c.* How easily we write such things! The main road from Nares Lucanae to Vibo, i.e. the Via Popilia, passed through a station called Interamnium, 100 miles from Nares Lucanae and twelve miles as the crow flies to the west of Thurii up the course of the River Sybaris (see Nissen, *Italische Landeskunde*, II, p. 918). If there is any evidence about communications between Thurii and the Via Popilia it has not come my way; but presumably C. could have taken a road eastwards from Interamnium which would have brought him to Thurii in 15 miles or so. Next morning he would have had to retrace his steps, or else, more likely, would have found a road south-west from Thurii leading along the River Crathis to Caprasia (Tarsia), 9 miles south of Interamnium along the Via Popilia. The crow flies 16 miles from Thurii to Caprasia; say 20 miles by road. We have therefore to believe that C., admittedly in some haste, chose for reasons unknown to travel an extra 26 miles or more in order to spend a night at Thurii. It does not seem probable.

(*c*) C.'s route from Nares Lucanae to Thurii needs closer inspection. On the evidence of ancient Itineraries the former is placed on the Via Popilia at a point 9 miles from the crossing of the River Silarus, which separates Campania from Lucania, near the modern Scorzo (see Philipp, *RE.* XVI. 1715). From there to Thurii was about 115 miles by road (see above). The villa in the Campus Atinas where Cicero stopped the night cannot be precisely located; this long valley begins 5 miles north of Atina and ends 14 miles south. C. may therefore have travelled nearly as far as Marcelliana, 28 miles from Nares Lucanae. That still leaves him with nearly 90 miles to negotiate in two days, all but the last 15 through the mountains of Lucania. This is excessive. A courier could average 45 miles or so without difficulty (cf. Friedländer, *Sittengeschichte Roms*[9], I, pp. 333 ff.); Caesar with a province to save could do much more. But C. was not a courier or a general in a hurry. With his retinue of friends, clients, and servants (cf. Plut. *Cic.* 31; App. *B.C.* II. 15; Cic. *Att.* 64 (III. 19). 1 fin.), and, doubtless, a sufficiency of baggage, he would do very well to average 35 miles over such terrain. When neither dawdling nor in special haste his normal rate might be about 25 miles (cf. Constans, *Correspondance*, III, p. 192, n. 3). A further element of improbability is the supposition that C. dispatched his letter from Thurii on the evening of 10 April, instead of the morning as was customary. It is fair to add that these difficulties are mitigated if Nares Lucanae be placed with Kiepert 14 miles further south, at the northern end of the Val di Diano; for C. might then have written III. 2 from the villa of his dream. But this location is discredited by Philipp (*l.c.*) and contrary to the evidence of the Itineraries. Another way out would be to change the date numeral in III. 2 from *VI* to *VII* or *VIII*, as Smith (p. 76) for a different reason wanted to do. But a theory which requires the alteration of *both* of our two given dates does not carry the signs of health.

(*d*) Smith (p. 74), followed by Sternkopf (p. 300), etc., reckons the distance from Thurii to Vibo as 85 miles. This is plainly an underestimate. Even as the crow flies, and he flies very straight in Calabria, it is 78 miles. By road, allow 20 miles from Thurii to Caprasia; Caprasia to Cosentia is 28 (Nissen, *op. cit.* p. 918), Cosentia to Vibo 57: total, 105 miles, much of it mountainous. That represents three days' journey, not two.

(*e*) I cannot believe that C. did not allow himself at least one day to recuperate on Sicca's farm. Arrived at his destination after eight or nine strenuous days on the road it is sheer inhumanity to drive him off the following morning on the 240 miles track to Brundisium. His situation did not require such unconscionable haste as that. Sicca was in no immediate danger, for Clodius' amended bill had not yet become law (III. 4 *ante diem rogationis*). Moreover, on Sternkopf's showing, C. had a good idea of its content before he left Thurii.

(*f*) Starting from Vibo on the 13th, Cicero has to reach Brundisium on the 17th (52 (III. 7). 1)—240 miles in five days (the distance has been abbreviated through the mis-reckoning noted above). That also is exorbitant.

An alternative scheme presented by de Benedetti must now be considered. She puts the letters in the order 1, 3, 4, 5, 2 instead of 1, 3, 2, 5, 4. III. 5 was written from Thurii on 6 April, as the MSS say, on the way back from Vibo. III. 2 from Nares Lucanae followed on the 8th. Gelzer in his biography (*RE.* VII A. 918. 47) agrees, except that he keeps the correction *IIII* for *VIII* in III. 5, thus putting III. 2 before III. 5. The contents of III. 2 do not favour these arrangements. Back at Nares Lucanae C. is not likely to have written to explain why he went to Vibo in the first place, and *itineris nostri* could scarcely mean *that* journey. But minor objections are superfluous in face of the monumental absurdity of finding him at Nares at all, wandering in north-west Lucania on his way from Vibo to Brundisium. De Benedetti's plea that the locality of Nares Lucanae is doubtful is mere subterfuge. Such doubt as may exist concerns a matter of 14 miles; and whether C. be assumed to have gone a little more or a little less than 100 miles north-west when he ought to have been going north-east is neither here nor there. Why should he? To make it easier, says Gelzer following de Benedetti's tentative footnote, for Atticus to meet him—as though C., who had already sent word that he was on his way to Brundisium (III. 4) and who did not even know whether Atticus had left Rome or not (III. 5), would have endured another 200 miles of mountain road on such an off-chance.

Between two unacceptable theories there is something to choose. The first, the conventional theory, is a concatenation of errors and improbabilities fit only for the dust-heap. The second breaks down at one point only, in respect of III. 2. Geography, which makes it certain that III. 2 precedes III. 4, makes it extremely probable that III. 4 precedes III. 5; for Thurii, which was not on the route from Nares to Vibo, doubtless was on the route from Vibo to Brundisium. But if III. 2 belongs to the journey down, and III. 5 to the journey up, one of the two dates must be wrong. Which one need be in no doubt. If C. wrote III. 2 from Nares Lucanae on 8 April he would not have covered the 435 miles *via* Vibo to Brundisium by 17 April. So *VI Id.* must be changed. Bearing in mind that C. has to find himself in Thurii on 6 April and to cover about 300 miles in order to get there *via* Vibo, one sees that he must have left Nares Lucanae in March; that *Id.* therefore should be *K(al).*

Such a change, says Smith in passing, is too violent to be thought of. Certainly it is less easy than that of *VIII* to *IIII*, but it is well within the range of the most conservative conceptions. Editors in general (I do not) assume the converse corruption in XV. 11. 1 and XVI. 2. 4; and they are right to assume it in XVI. 10. 1, where the Tornesianus, independently attested by two witnesses, had *VI Kal.* and the context requires *VII Id.*[1] Here *K(al).* may have become *Id.* simply under the influence of the similar date in III. 5.

Let us see how this works, noting first that, apart from the inversions of III. 2 and III. 3, the letters are now back in the order of the MSS, which in this part of the correspondence is as a rule chronologically correct. Starting from Nares Lucanae on 27 March C. spends the night in the Campus Atinas, say 26 miles down the road. That left him 168 miles to go, more or less according to the whereabouts of Sicca's farm. Admittedly in haste, he may have averaged $33\frac{1}{2}$ miles a day even through this difficult country and gained Vibo on 1 April. Setting out again on the 3rd, he reaches Thurii on the 5th, 105 miles in three days. We could give him an extra day on the farm by changing the date in III. 2 to *VII K(al).* instead of *VI K(al).*; palaeographically there is little to choose.

Setting out from Thurii on 6 April C. might be expected to arrive at Tarentum, 90 miles away, on the 8th. He is generally assumed to have arrived on the 16th, since his next letter, III. 6, is dated *XIIII Kal. Mai. de agro Tarentino*. If that be so, accident or sickness will have delayed him. But there is nothing in III. 6 to show that he spent only one night near Tarentum. His remark that he had hoped to have seen Atticus at either Tarentum or Brundisium is consistent with a longer stay. He reached Brundisium next day—the distance may have been appreciably less than the full 45 miles between the two towns since he travelled not from Tarentum itself but from somewhere in the *ager Tarentinus*—and stayed (from the 17th to the 29th) in the *horti* of M. Laenius Flaccus on the outskirts of Brundisium (*Fam.* XIV. 4. 2; *Planc.* 97). Nothing forbids us to suppose that he spent a week near Tarentum before continuing his journey.

To go back to 27 March. Allowing three days for the 87 miles between Teanum Sidicinum and Nares Lucanae, we may suppose that III. 3 was written on the 24th. C. may therefore have started out from Formiae on the 23rd or from Arpinum on the 22nd, dates agreeable to the view that he left Rome in the third week of March.

I hope it is not too much to claim that *res* and *ratio* demand the textual change I have proposed; also that they have been known to demand harder things in voices less peremptory.

[1] Sjögren puts Sternkopf's *Kal.* for *Non.* in his text at 65 (III. 20). 3. Editors make the same substitution in *Fam.* XV. 3. 1 (wrongly).

APPENDIX II

THE DATES OF LETTERS 77–86 (IV. 4*a*–12)

The dating of these letters has most recently been discussed by Miss L. R. Taylor in a valuable paper: *Cl. Phil.* 44 (1949), pp. 217–21.

77 (IV. 7) is definitely the earliest. It clearly comes shortly after *Q.Fr.* II. 6 of 9 April 56, which was written *en route* to Arpinum, where C. expected to spend 11–15 April (§4).

Their contents show that 78–80 (IV. 4*a*, 8, 5) were written from the same place and close in time. That the place was Antium (79 (IV. 8). 1) and the year 56 is undisputed. The month is indicated only by C.'s reference in 80 (IV. 5). 1 to a 'palinode' which he had composed to conciliate Caesar. The only extant composition with which that can plausibly be identified is the speech *de Provinciis Consularibus*, and despite the arguments of C. Saunders (*Cl. Phil.* 14 (1919), pp. 201 ff.) and T. Rice Holmes (*Cl. Quart.* 14 (1920), pp. 21 ff.) to the contrary, I think that this remains the best answer. The hypothesis of a private letter to Caesar or Pompey is discountenanced by C.'s statement in §2 that he had wanted to tie himself down irrevocably to his new alliance with the dynasts; a private letter would hardly involve such *necessitas. mea* too ('my writings') suggests a composition for public reading—taken of private letters (cf. 134 (VII. 11). 5 *legens tua*) the sentiment is absurd—and ἀποθεώσει, if correct, a eulogy such as the *de Provinciis* actually contains. As for the term 'palinode', the effect of the speech upon contemporaries must have been just that. A glowing, public panegyric of Caesar's achievements from C.'s lips could only present itself as a *volte-face*, as C. himself practically admits in the apologetic §§ 18 ff. A public letter, such as C. thought of writing to Caesar in 45 (281 (XII. 40). 2, etc.) is not absolutely outside possibility, but there is not a particle of evidence that C. ever wrote or published such a letter in 56, whereas we know that he published the *de Provinciis*. The most plausible alternative is that proposed by J. P. V. D. Balsdon (*Journ. Rom. Stud.* 52 (1962), p. 139): an earlier speech (or pair of speeches) in favour of Caesar's requests for a financial grant (*stipendium*) and ten *legati* (cf. *Prov. Cons.* 28; *Balb.* 61). C.'s promise to go further if Caesar received his offering gladly (§2) is in favour of this, but clearly not conclusively so. Against it is again the absence of evidence for such a publication.

The *de Provinciis* belonged to late May or June, being subsequent to Q. Cicero's return from Sardinia (cf. *Fam.* I. 9. 10), which itself was some time after 16 May (cf. *Q.Fr.* II. 7. 2). The three letters therefore may plausibly be placed in mid- to late June (see below). 78 (IV. 4*a*) is certainly the first. C. asks Atticus to send him two library assistants, whose operations are praised in both the other letters. He also hopes that Atticus will pay him a visit and mentions his recently purchased gladiators. 80 (IV. 5) is generally placed before 79 (IV. 8) because its final paragraph is supposed to renew the topic of Atticus' visit, which does not appear in 79 (IV. 8); but this is a mistake (see notes). Otherwise the evidence points to the reverse order. Both letters are replies to letters received from Atticus. The mention of gladiators at the end of 79 (IV. 8) suggests that Atticus had recurred to the subject, in which case *his* letter probably answered 78 (IV. 4*a*) and will have been carried to Antium by the two *librarii*. The work of these *librarii* is warmly commended in 79 (IV. 8). 2, but nothing is said about sending them back; whereas the wording of 80 (IV. 5) *bibliothecam*...*laudes* indicates that they had done their job and were going home, probably carrying C.'s letter.

81 (IV. 12), formerly assigned to April–June 55, belongs according to Miss Taylor (*l.c.* p. 220) to some later month before November 55. Her argument, which is not very clear to me, seems to depend on the assumption that the letter was written at Tusculum, for which there is no evidence. The words *de re †Halimeti† vehementer Anti egi* do not disprove Antium as the place of origin (cf. 26 (II. 6). 1 *libris*...*quorum habeo Anti festivam copiam*), nor on the other hand do they warrant Constans' deduction (II, p. 121) that Cicero wrote from Antium or on his way from Antium to Rome. Neither the year nor the month can be definitely established. C. writes of going 'home' (*domum*) and asks Atticus and Pilia to dinner *apud me*, but it cannot be inferred that his house on the Palatine was already habitable; these expressions may just as well refer to the house which he had doubtless rented in Rome (cf. *Q.Fr.* II. 3. 7) in 57–56. His proposed journey to Larinum to attend an auction suggests a month between March and October; and the reference to dinner at Crassipes' *horti* ties up neatly with 80 (IV. 5). 4. For the latter reason, and not for those he gives (II, p. 122), I follow Constans in associating this letter with the Antium group of 56. The Kalends on which C. intends to return to Rome will therefore be those of July (see above).

82 (IV. 8*a*), evidently from a place near Rome (§4), Tusculum therefore or Antium, is assigned to the autumn of 56 by a reference to Domitius Ahenobarbus' candidature for the Consulship. The games in §1 are likely to be the *ludi plebeii* (4–17 November) rather than the *Romani* (4–19 September): cf. Constans, II, pp. 122 f. C. seems to have written shortly before 18 November, which was added to the games this year (cf. §1 *statim fac ut sciam*, taken in conjunction with the following sentence).

84 (IV. 10) and 85 (IV. 9) date themselves. The former was written at Cumae on 22 April 55, the latter at Naples or *en route* to Pompeii on

27 April. Miss Taylor convincingly assigns 83 (IV. 6), previously associated with the Antium group of 56, to this Campanian sojourn, placing it about eight days before 85 (IV. 9). Her reasons are: (*a*) the striking difference of mood between 80 (IV. 5—lively and down-to-earth) and 83 (IV. 6—mournful, despairing); (*b*) the arrival of four letters from Atticus simultaneously points to somewhere further from Rome than Antium; (*c*) the lack of community of topics, apart from §4 *domum nostram...invisas*, the wording of which goes better with 84 (IV. 10). 2 and 85 (IV. 9). 2; (*d*) 85 (IV. 9) with its references to Lucceius and the Palatine house reads like a reply to Atticus' reply to 83 (IV. 6).

86 (IV. 11) has generally been assigned to 26 April, the day on which C. left Cumae for Naples (85 (IV. 9). 2). But why then does he say nothing about leaving? Clearly he was settled for the moment, deep in literary work (§2). T.–P.'s appendix (II, pp. 269 f.) illustrates the hopeless complications involved in the traditional date. Miss Taylor (*l.c.* pp. 218 f.) may be followed in making June the time and Tusculum (or possibly Antium) the place. June is indicated by references to a gladiator show such as candidates were apt to give before the elections, and a place near Rome by Pompey's reported statement that he expected Crassus at his Albanum the next day, the 27th, and other indications.

APPENDIX III

LETTER 82 (IV. 8*a*) 2

82 (IV. 8*a*). 2 quid enim hoc miserius quam eum qui tot annos quot habet designatus consul fuerit fieri consulem non posse, praesertim cum aut solus aut certe non plus quam cum altero petat?

Why does C. say that L. Domitius Ahenobarbus was standing by himself for the Consulship of 55, or, at any rate, had only a single competitor? This is Tyrrell–Purser's answer: 'Domitius was the only plebeian candidate except Pompey (Crassus being a patrician); but Cicero hesitates whether he can rightly call Pompey a candidate at all, since he was not elected, but appointed colleague of Crassus after an *interregnum*. By the Licinian constitution one of the consuls was necessarily a plebeian. Others think that Pompey is not here referred to, but some plebeian rival of whom no further mention is made....' Useless to ask where they found the notion that 'Pompey was not elected, but appointed colleague of Crassus', or how they supposed that C. was able to foresee this remarkable procedure. But the doctrine that Crassus was a patrician belongs to the common stock of blunders, faithfully transmitted from Paulus Manutius down to Constans; even outside the petrified forest of Ciceronian editions one meets it occasionally, as (implicitly) in Eduard Meyer's *Caesars Monarchie* (p. 153, n. 1). Manutius was further mistaken (though less categorically so than some of his successors) in his idea that Crassus' supposed patrician status would explain C.'s language here, as though patrician candidates for the Consulship did not compete with plebeian under 'the Licinian constitution'. They did, of course.

From the sources listed by Broughton (p. 214), mainly Dio and Plutarch, the outline of what happened is fairly clear. Pompey and Crassus did not present themselves as candidates by the proper date. However, when their intention to take the Consulships for 55, as arranged at Luca, had become manifest, the candidates already in the field, except Domitius, withdrew. Since it was plain that the Consuls of 56, especially Lentulus Marcellinus, would refuse to accept the belated candidature of the two 'Triumvirs', as they were legally entitled to do, these put up a Tribune to veto the holding of the elections in 56. At elections held in January 55 under an *interrex* Pompey and Crassus were duly returned, after their strong men had beaten Domitius and his backer M. Cato off the *campus*.

If C. in November had considered Pompey and Crassus as candidates

Domitius would have had at least two competitors. Evidently he did not, and for the good reason that their candidatures had not been formally announced and accepted. The competitor about whom he is doubtful is an unknown. He must have been one of those who withdrew—Dio mentions (XXXIX. 27. 2) that the 'Triumvirs' had been supporting other candidates, ἑτέροις τισὶ πρότερον συναγωνιζόμενοι, before they came forward themselves—but when C. wrote there was still some uncertainty about this person's withdrawal. He may have been a plebeian or he may not.

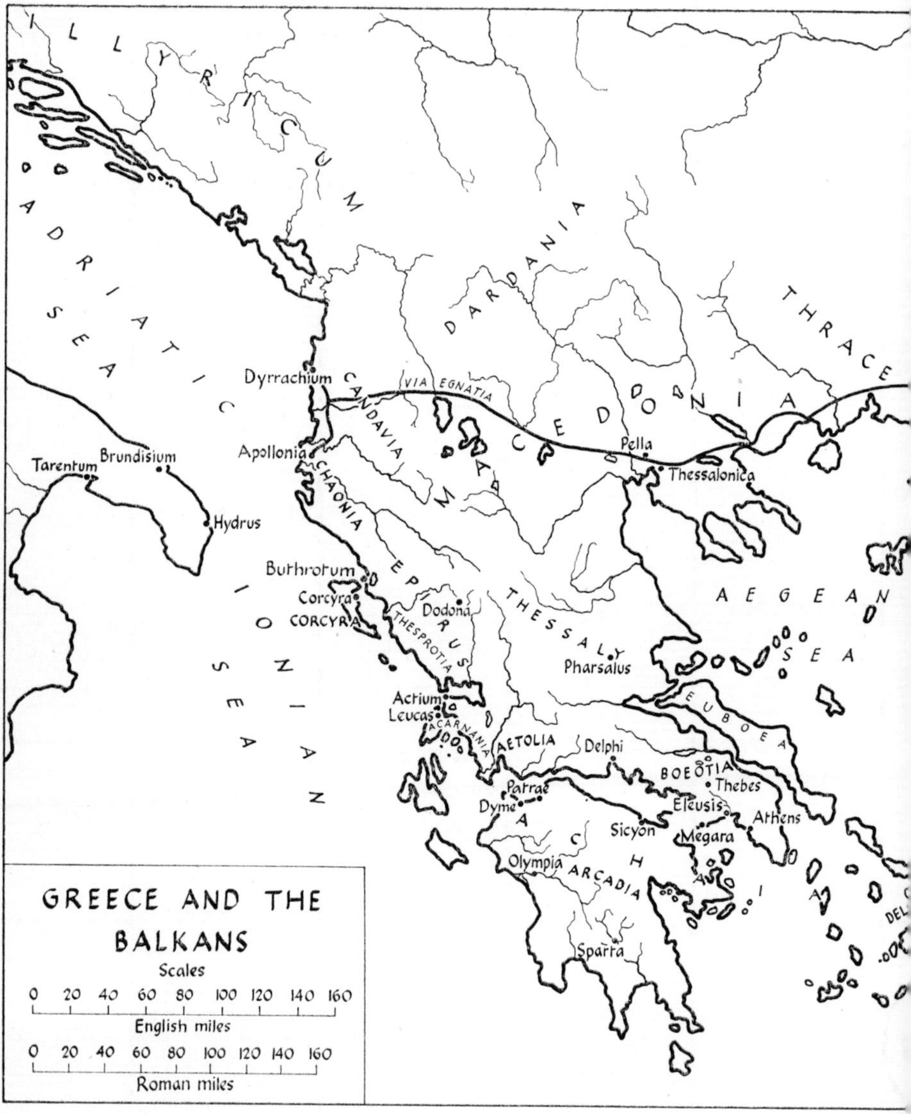

GREECE AND THE BALKANS

CONCORDANCE

This Edition	*Vulg.*	*Vulg.*	*This Edition*
46	III. 1	III. 1	46
47	III. 3	III. 2	48
48	III. 2	III. 3	47
49	III. 4	III. 4	49
50	III. 5	III. 5	50
51	III. 6	III. 6	51
52	III. 7	III. 7	52
53	III. 8	III. 8	53
54	III. 9	III. 9	54
55	III. 10	III. 10	55
56	III. 11	III. 11	56
57	III. 12	III. 12	57
58	III. 14	III. 13	59
59	III. 13	III. 14	58
60	III. 15	III. 15	60
61	III. 16	III. 16	61
62	III. 17	III. 17	62
63	III. 18	III. 18	63
64	III. 19	III. 19	64
65	III. 20	III. 20	65
66	III. 21	III. 21	66
67	III. 22	III. 22	67
68	III. 23	III. 23	68
69	III. 24	III. 24	69
70	III. 25	III. 25	70
71	III. 26	III. 26	71
72	III. 27	III. 27	72
73	IV. 1	IV. 1	73
74	IV. 2	IV. 2	74
75	IV. 3	IV. 3	75
76	IV. 4	IV. 4	76
77	IV. 7	IV. 4*a*	78
78	IV. 4*a*	IV. 5	80
79	IV. 8	IV. 6	83
80	IV. 5	IV. 7	77
81	IV. 12	IV. 8	79
82	IV. 8*a*	IV. 8*a*	82
83	IV. 6	IV. 9	85
84	IV. 10	IV. 10	84
85	IV. 9	IV. 11	86

CONCORDANCE

This Edition	*Vulg.*	*Vulg.*	*This Edition*
86	IV. 11	IV. 12	81
87	IV. 13	IV. 13	87
88	IV. 14	IV. 14	88
89	IV. 16	IV. 15	90
90	IV. 15	IV. 16	89
91	IV. 17	IV. 17	91
92	IV. 18	IV. 18	92
93	IV. 19	IV. 19	93

INDICES

To Commentary and Appendices only. References to Commentary are by letter (serial number of this edition), paragraph and line of lemma.

Further indices are planned to appear in a separate volume at the end of the series.

I. INDEX NOMINVM

II. INDEX VERBORVM

A. Latinorum

B. Graecorum

III. INDEX RERVM